Green history

Green history

A reader in environmental literature, philosophy and politics

Derek Wall

London and New York

8-23-96

First published 1994
by Routledge
11 New Fetter Lane, London EC4P 4EE
Simultaneously published in the USA and Canada
by Routledge
29 West 35th Street, New York, NY 10001

Typeset by Solidus (Bristol) Ltd, England
Printed in Great Britain by Clays Ltd, St Ives PLC

British Library Cataloguing in Publication Data
A catalogue reference for this book is available from the British Library

Library of Congress Cataloging in Publication Data

Wall, Derek.
Green history : an anthology of environmental literature.
philosophy, and politics / Derek Wall.
p.   cm.
Includes bibliographical references and index.
1. Green movement—History.   2. Environmental sciences–
Philosophy–History.   3. Environmental policy–History.   I. Title.
GE195.W35   1993
363.7′0525′09–dc20   93-28232

ISBN 0–415–07924–1 Hb
ISBN 0–415–07925–X Pb

To Mark B. and Mark S.

Contents

Contents

Contents

Contents

Acknowledgements

Permission to reprint copyright material from the following sources is gratefully acknowledged. Extract from 'The discovery of America' by Paul S. Martin, pp. 969, 972 of *Science*, volume 179, 3 September 1973, by permission of Paul S. Martin and the American Association for the Advancement of Science, copyright 1973 by the AAAS. Extract from *Ecology in Ancient Civilizations* by J. Donald Hughes, reprinted by permission of Professor J. Donald Hughes. Extract from *The White Goddess* by Robert Graves, reprinted by permission of A. P. Watt Ltd on behalf of the Trustees of the Robert Graves Copyright Trust. Extract from J. Smuts, 'Introduction' to J. Bews, *Human Ecology*, Natal: Natal University (1935), reprinted by kind permission of Mr J. C. Smuts and Natal University. L. Mumford, *The Culture of Cities*, London: Secker and Warburg, pp. 300–4. Extracts from *Point Counter Point* and *Island* by Aldous Huxley, published by Chatto & Windus, reprinted by permission of Mrs Laura Huxley and Random Century Group. Extract from *Science and Society in Ancient China* by Joseph Needham, reprinted by permission of Cambridge University Press. Extract from *The Road to Wigan Pier* by George Orwell, reprinted by permission of the estate of the late Sonia Brownwell Orwell and Martin Secker & Warburg Ltd. Extract from *Hildegard of Bingen*, edited by F. Bowie and O. Davies, reprinted by permission of the Society for Promoting Christian Knowledge. Extract from *Small is Beautiful* by E. F. Schumacher, published by Hutchinson, reprinted by permission of Random Century Group. Extract from *Capitalism, Socialism and the Environment* by Hugh Stretton, reprinted by permission of Cambridge University Press. Extract from *The Eclipse of Reason* by Max Horkheimer reprinted by permission of The Continuum Publishing Company, copyright 1947 Oxford University Press, New York, new material copyright 1974 The Seabury Press. Extract from *Man and Nature: The Spiritual Crisis in Modern Man* by Syed Nasr reprinted by permission of Unwin Hyman, an imprint of HarperCollins Publishers Limited. The extract reprinted on pages 186–189 is from *Women and Nature* by Susan Griffin, first published in Great Britain by The Women's Press Ltd, 1984, 34 Great Sutton Street, London EC1 0DX, copyright Susan Griffin 1984, c/o Artellus Ltd, 30 Dorset House, Gloucester Place, London NW1 5AD. 'Were the Greeks Green' by L. Goodison reproduced by kind permission of *Greening*

Acknowledgements

the Planet magazine. 'Longing the die of old age' from *Living by the Word* by Alice Walker reprinted by permission of Harcourt Brace and Company and The Women's Press Ltd, copyright 1983 Alice Walker. Extract from 'Nobody was supposed to survive' from *Living by the Word* by Alice Walker reprinted by permission of Harcourt Brace and Company and The Women's Press Ltd, copyright 1983 Alice Walker. Selected excerpt from *The Spiral Dance* by Starhawk copyright 1979 by Miriam Simos, reprinted by permission of Harper-Collins, Publishers, Inc. I would particularly like to thank Mark Lamont Brown for reading drafts of this project. Completion would have been equally impossible without the help of Karen Bell, Ian Coates, Scott Cooper and Mark Simpson or the support of my editor, Tristan Palmer, Jacqueline Cohen, Gideon Kossoff and Dr Ian Welsh read the manuscript and offered valuable criticism. Thanks must go also to the staff of the following libraries: Bath University, Bradford University, Bristol City Reference, Bristol University, the Commonwheal Collection at Bradford University, Sussex University, the University of London and the University of the West of England. Finally the long-suffering staff of the British Library deserve my gratitude.

Introduction

For most of us, even the committed activist, the Green movement has no history. Worries about environmental destruction seem very modern. Acid rain, the greenhouse effect and ozone depletion are concerns of the last twenty years, especially of the last four or five. Our great-grandparents never had to worry about nuclear waste. Prior to Hiroshima and Nagaski, a Green movement would, in one sense, have been impossible. Robert Oppenheimer, leader of the team that unleashed the power of the atom, watched the first test detonation of an atomic bomb in the New Mexico desert, first with awe and then with mounting horror. He recalled a phrase from the Hindu epic, the *Bhagavad-Gita*, 'I am become Death, the shatterer of worlds' (Worster 1991: 339). The verse and its implications echo through the years since. In a dozen ways, or more, we can now destroy life on a world-wide scale. Global warming, degraded seas, an ever rising tide of low-level radiation, species loss, high-technology warfare, the risks of biotechnology and even electromagnetic pollution, all come to mind as potentially terminal threats to Planet Earth. The contemporary Green movement was born in response to the feeling that we have 'become death'.

To many, this movement is '. . . like a stranger who had just blown into town, . . . a presence without a past' (Worster 1991: xiii). Green activists, their opponents and the ever watchful news media proclaim the novelty of an ecological outlook. Greens of all shades seem, often with unintended irony, to echo the words of that arch-architect of twentieth-century capitalism, Henry Ford. 'History', more often than not, for radical environmentalists, eagerly rejecting the polluting legacy of the last hundred years, 'is bunk'. Dobson (1990: 182) argues that the Green movement presents its self as 'something uniquely fresh and novel'. Capra and Spretnak argue that the world has been dominated for centuries by a 'paradigm' of domination, reductionist science and masculine values hostile to the natural world. For them 'Green politics' represents 'the political manifestation of the cultural shift' to a new 'paradigm'; they conclude that 'What we need is a new dimension of politics altogether. Green politics offers such a dimension, a politics that is neither left nor right but in front' (1986: xvii–xviii). Although other Greens may, in contrast to Capra

1

and Spretnak, place themselves rather more firmly on the left, this theme of an entirely new philosophy is repeated again and again in pamphlets, political broadcasts and Green texts.

I argue not only that the Greens have a history but that both they and wider society would do well to learn from it. Political movements can grow by drawing upon the resources of the past. For good or for ill, the past can be used as a flag to lead supporters into a new future. Given that communists and conservatives, regionalists and even feminists gain strength by celebrating their heritage and their heroic figures, why should Greens act differently? But even if they reject such instrumental and propagandist use of their roots, Greens have other gains to make from their past. It is virtually a cliché that, by tracing the origins and social context of ideas, it becomes easier to understand their practical implications and significance. The issues explored by modern environmentalists are not new, as Martinez-Alier notes. 'Human relations with the environment have a history, and the perception of such relations is also historical' (1990: xiv). Even if the Green movement endorses in its brand-name innocence Gramsci's schoolmasterly dictum that 'History teaches but has no pupils', there is no reason why others should buttress his pessimism with such ignorance. To understand the growth and nature of environmentalism we must study its roots. Thomas, in carrying out such a task, observes,

> Nowadays one cannot open a newspaper without encountering some impassioned debate about culling grey seals or cutting down trees in Hampton Court or saving an endangered species of wild animals. But to understand these present-day sensibilities we must go back to the early modern period.
>
> (1984: 15)

The history of Green movements and the wider study of the relations between humanity and nature should be of interest to all who seek a solution to the problems of pollution and environmental degradation. One suspects that it would do us no harm to examine how our ancestors responded (or failed to respond) to soil erosion that robbed land of fertility, air pollution that killed children, or heavy metal contamination that sent citizens mad. The environmental past may, sometimes, indicate useful parallels with the environmental present.

Hazards of pollution, deforestation, land degradation and chemical food adulteration have dogged humanity, to a greater or lesser degree, for most of its existence. Environmental archaeologists have created a science out of the study of catastrophes suffered by ancient civilizations and prehistoric peoples (see Chapters One and Two). Heavy metal pollution may, it has been suggested, have been a factor in 'the fall of Rome': cooking vessels, plates and pipes were all made using lead or alloys of the metal. The excesses of Caligula and Nero may have been due to their fondness for acidic fish sauces that dissolved the lead from the imperial tableware, into the imperial brain and bloodstream (Nriagu 1983). Numerous ancient societies have collapsed because of environmental degradation. The builders of Avebury and Stonehenge seem likely to have caused massive deforestation, leading to soil erosion, climatic change and probable famine. The Mayan pyramid builders may have caused their own demise in a similar fashion. Over-zealous irrigation schemes that drew salt into the soil hastened the collapse of Sumerian society and possibly that of the Indus valley. Plato, showing a keen understanding of basic ecological principles, described soil erosion in fourth-century BC Attica:

> In comparison with what then was, there remain only the bones of the wasted
> body ... all the richer and softer parts of the soil having fallen away, and the
> mere skeleton of the land being left ... the land reaped the benefit of the
> annual rainfall, not as now losing the water which flows off the bare earth into
> the sea.

He blamed overgrazing by goats – a common cause of soil loss in today's sub-Saharan
Africa (see extract in Chapter Two). A few centuries later a Roman writer, Lucretius,
concluded that the Earth was exhausted, noting that 'ancient ploughmen' already shook
their heads and sighed over the easier, more productive fields that their fathers had tilled.
Caesar remarked on the hedge-laying activities of his enemies, the Nervii, during the
Gallic wars (Allaby 1975: 224). Hedges were and remain an important means of preventing
soil erosion.

Green movements and themes, like environmental problems, also have a surprisingly
long past. In the same way that Mrs Thatcher looked to Adam Smith and Samuel Smiles,
or modern Marxists look beyond their bearded mentor to the British economists, French
socialists and Germany philosophers, from whom he drew his inspiration, the ideas of
modern political ecologists such as Bookchin, Foreman, Porritt and the late Petra Kelly
have roots in earlier thinkers. For example, you cannot talk of Foreman and deep ecologists without noting the inspiration of John Muir, Henry David Thoreau and the native
American 'Indian'. Kelly looked to a century of women's agitation for peace, to E. F.
Schumacher, the *Small is Beautiful* economist, and to Gandhi. Bookchin seeks to transform
Hegel, drawing strongly on his dialectic philosophy, and was enlightened by the ecological
anarchism of Prince Peter Kropotkin. Porritt draws, in part, upon R. H. Tawney and via
him taps an older tradition of Guild Socialism and decentralized pastoral politics. If we
take the science of ecology, as an essential of the Green outlook, in understanding its
development, we must examine not just Haeckel, who provided the child with a name, but
earlier figures, including Darwin, Gilbert White and John Ray (Chapter Eight). Ecological
science, in turn, was strongly influenced by a philosophy of holism, from which it cannot
be divided. This philosophy was drawn, variously and eclectically, as we shall see
(Chapter Seven) not only via Hegel, Heidegger or Spinoza or as Capra suggests from the
'New Physics' but from quasi-religious and occult sources – Buddhism, Gnosticism, the
Kabbala, Neoplatonism, Sufism and Taoism – and synthesized in a European context by
Blake, Goethe and other Romantics. Could Haeckel have given ecology its name without
being familiar with his countryman Goethe's holism? The utopian ruralist politics of
nineteenth-century British radicals such as William Cobbett draws upon earlier practitioners
and influences modern Green politics (Chapter Thirteen). Other Green themes such as
decentralism (via Kropotkin and Godwin) and animal rights (from Shelley and Pythagorus) have equally diverse and ancient origins. Many current debates within the ecology
movement such as the dispute between 'deep' and 'social' ecology can, equally, be prefigured in the past. By doing more than note the contours of earlier environmental debate,
modern environmentalists of all varieties might save themselves the need laboriously to
reinvent numerous intellectual wheels.

Charting these currents of prefigurative debate is no easy task. Green history comprises
a huge and ill defined territory. Drawing a boundary around the subject and listing its
contents is a difficult enough task. Examining the significance of such elements, placing

them in a living social context and illustrating their interrelation is rather harder. The study of past Green movements and of wider environmental history is a relatively new discipline. Martinez-Alier, for example, notes that the European Society for Environmental History held its inaugural meeting as recently as 1988 (1990: xv). With the honourable exceptions of environmental archaeology and the kind of landscape history practised from the 1930s by G. M. Trevelyan, it has been rare to find studies of the past that examine social change in an ecological context. Green history that specifically charts the origins of Green movements is even newer. Sometimes not a lack of research but an abundance of contradictory discourse may give rise to uncertainty, with the would-be Green historian entering a multi-disciplinary minefield. To understand the origins of Green thinking we need to enter into debates about the nature and importance of such complex (and often obscure) currents as distributionism, Europe's occult underground, the Frankfurt school, Haeckel's monism and Romanticism.

The term 'Green' is itself, open to ambiguous and shifting interpretation. George Bush and Margaret Thatcher have been labelled 'Green', we have 'Green' capitalism and consumerism. The term has become increasingly vague, at worst denoting a self-defined sign for those opposing environmental sin with little practical commitment and even less intellectual rigour. Gould, in, perhaps the first study of *Early Green Politics*, writes:

> It is not possible, however, to be very precise in the definition of the two imperatives in the title of this study. They did not come from one centre or convey one set of ideas. The wide range of ideas which they embrace makes clarity of definition even more difficult than in the case of other creations of the human mind.

(1988: 156)

Even if we reject this pessimism as excessive, a difficult question remains: what do we mean by 'Green'? At the very least, it seems obvious that there is a gulf of difference between the Green parties of the 1980s and, to take an extreme example, the former US President George Bush! As a starting point, I have taken Green/ecological politics/radical environmentalism to be defined, as essentially interchangeable terms, by the programmes and philosophy of the European Green parties formed in the late 1970s and early 1980s. Such an outlook is described by active Greens (Capra and Spretnak 1986; Kelly 1984; Kemp and Wall 1990; Porritt 1984), in party publications and by independent observers. Despite areas of ambiguity and difference, a number of key features of a 'Green' package, can be found in all these sources. The 'Green' package is based on environmental concern but does not merely state that the environment should be preserved for human benefit. Non-human nature is given philosophical status outside our own interests. Destroying the rain forests and polluting the seas may hurt human society, but even if such acts brought unalloyed benefit to our species they would be opposed by Greens. This attitude of 'deep ecology' gives not just the whale and the antelope status but also the forest, the ocean and even, on occasions, mountains an ethical standing. The leftist intellectual and former East German dissident Rudolf Bahro left the German Green Party over its, in his view, contradictory failure to speak out against all forms of animal exploitation, including vivesection for vital medical research (1986: 196–209). While, with the exception of some feminist spirituality, most Greens reject the notion of nature spirits or a conscious Earth Goddess, all see nature as alive, vital and worthy of reverence. Thus Greens are strongly anti-anthropocentric and ultimately perhaps rather pagan in outlook.

Despite this pagan love of all life and somewhat mystical undertones, the Green 'package', as we have already noted, is virtually unthinkable without the contribution of scientific ecology. Ecologists warning of environmental catastrophe helped launch the earliest of contemporary Green parties and movements. Investigations over the last 150 years have shown, research paper by research paper, the penalties we pay for disrupting the environment. By 1880 scientists had developed the concept of the greenhouse effect to illustrate how fluctuations in CO_2 had led to climatic change in the past. By the 1930s soil erosion had become a recognizable problem. In 1946 Sir George Stapledon started to write on the topic of *Human Ecology*. And in 1959 Dr David Price of the US Public Health Service examined the risk of an unprecedented catastrophe:

> We all live under the haunting fear that something may corrupt the environment to the point where man joins the dinosaurs as an obsolete form of life ...
> And what makes these thoughts all the more disturbing is the knowledge that our fate could perhaps be sealed twenty or more years before the development of symptoms.
>
> (Quoted by Barr 1970: 42)

In 1962, under the pseudonym Lewis Herber, the Green anarchist and academic Murray Bookchin published a critique of the 'full range of modern technology's incidental effects: polluted air, food with pesticide residues, milk contaminated by strontium 90, intolerable living conditions in cities, water not fit to drink, diets of chemical additives ...' (Cohen 1984: 292). Six months later Rachel Carson published the better known *Silent Spring* and by 1972 influential reports like *Blueprint for Survival* and *The Limits to Growth*, were arguing that infinite growth in human numbers, refuse and pollution, would inevitably lead to crisis. The Boston-based biologist Professor Barry Commoner stood as a presidential candidate on a radical Green programme during the 1980s, as did our final contributor, Dumont, in the French election of 1974. Scientific ecologists also helped to launch the UK Ecology Party in the 1970s.

Although the term was coined only in 1869, 'the idea of ecology is much older than the name' (Worster 1991: xiv). Ecology and economics have a common origin in the Greek term *oikos*, meaning 'home'. *Oeconomy* was first used to refer to household management in 1530, and in 1658 the naturalist Sir Kenelm Digby used it to discuss the 'oeconomy of nature.' In the late seventeenth century authors like John Ray in his *The Wisdom of God manifested in the Works of Creation*, recognized and attributed the beauty and stability of nature to God's hand. In similar vein the Green-tinged cleric William Paley wrote his *Natural Theology* in 1802. More significantly Carl von Linné, Linnaeus, the father of botany and zoology, published *The Oeconomy of Nature* in 1749. Such an 'Oeconomy' was already giving rise to conservation concerns. Nicholas Collin implored members of the American Philosophical Society in 1793 to support protection for certain bird species until it was discovered 'what part is assigned to them in the oeconomy of nature'. Curiosity about the environment was thus a rising tide, and a few years later Humbolt decided that he must voyage around the world to discover how nature's balance was achieved, 'In short, I must find out about the harmony of nature' (Worster 1991: 133). But it was another voyage, that of the *Beagle*, that was to launch ecology as a science. Darwin's theory of evolution was vital to the creation of ecology. If, as he claimed, humans had evolved from apes, perhaps we had a moral obligation to treat related creatures with a little more 'humanity'. The novelist Thomas Hardy, who had sympathies with the anti-vivisection movement, also argued that 'the most far-reaching consequence of the establishment of the common origin of species is ethical' (1930: 141–2). Darwin saw other species as 'our fellow brethren' and was critical of animal abuse (Nash 1989: 43). He was also deeply concerned with the workings of nature and did much to put the science of ecology on its feet. The publication of *On the Origin of Species* coincided with Haeckel's work on ecology. Parallel research on the part of another Englishman, Sir George Mirvart, led briefly to the use of a rival term, 'Hexicology'. Mirvart, like Haeckel and Darwin, was able to draw practical conclusions from his studies:

> Let a new land be discovered with a peculiar fauna and flora full of scientific interest, and straitaway the European purposely introduces his thistles, his sparrows, his rabbits or his goats, and the harmonious balance which has resulted from the organic interplay of ages is at once destroyed. Downright evil is often the result. Forests are recklessly felled, and arid rainless wastes or dismal fever-laden swamps ensue.
>
> (Syer 1971: 13–14)

Scientific ecology, with its appreciation of webs and relationships that link species, including humanity, gives some practical cement to a Green belief in 'holism'. Ecology alone does not equal ecological politics. Worster notes a division between two traditions in the science of ecology: the Arcadian and the Imperialist (1991: xi). The Imperialist ecologist uses the subject to discover better ways of 'managing' nature for human benefit, the Arcadian advocates the 'deep ecology' approach of giving non-human life independent ethical status. The Imperialist seeks to exploit, the Arcadian to live in harmony. Engels shows that it is possible to be concerned about the environment and aware of ecology without being a 'Green' when he argues:

> with every day that passes we are learning to understand these laws more correctly, and getting to know both the more immediate and the more remote consequences of our interference with the traditional course of nature. In parti- cular, after the mighty advances of the natural sciences in the present century, we are placed more and more in a position where we can know, and hence control, even the most remote natural consequences at least of our most ordi- nary productive activity.
>
> (Parsons 1977: 180)

Such views might be usefully contrasted with those of his near contemporary, the US naturalist John Muir:

> Nature's object in making animals and plants might possibly be first of all the happiness of each one of them, not the creation of all for the happiness of one. Why ought man to value himself as more than an infinitely small composing unit of the one great unit of creation?
>
> (Devall and Sessions 1985: 104)

Thus Engels praises the science as a means of improving human control, Muir uses it to suggest that humanity is merely one amongst many interwoven but equal forms of life.

Introduction

Worster writes that 'the present corpus of ecological thought is a conglomeration of all its pasts, like a man who has lived many lives and forgotten none of them', noting further that 'It was precisely against that easy identification of ecology with a particular moral philosophy that I wrote this book' (1991: xvii).

'Arcadian' Greens reject economic growth, argue that nature knows of no infinite expansion and cannot cope with ever increasing human demands upon it. Such rejection has a history. In his novel *Point Counter Point*, written in 1928, Aldous Huxley allowed one of his characters to speak for him on the subject of 'Progress':

> Lord Edward started at the word. It touched a trigger.... 'Progress! ... You politicians are always talking about it. As though it were going to last. Indefinitely. More motors, more babies, more food, more advertising, more money, more everything, for ever. You ought to take a few lessons in my subject. Physical biology. Progress, indeed!'

Such a concept of a 'stationary' economic cycle was praised by John Stuart Mill in his *Principles of Political Economy*, published in 1848. He did not regard

> the stationary state of capital and wealth with the unaffected aversion so generally manifested towards it.... I confess I am not charmed with the ideal of life held out by those who think that the normal state of human beings is that of struggling to get on; that the trampling, crushing, elbowing, and treading on each other's heels, which form the existing type of social life, are the most desirable lot ...

A critique of growth, environmental concern, scientific ecology, philosophical holism and the granting of status to non-human nature is vital to our definition but these inter-linked attitudes do not, in themselves, supply us with a complete picture of the Green package. In seeking to solve environmental problems Greens have been forced to consider human affairs and embrace a set of political, economic and cultural principles. Any description of 'Green' demands an exploration of approaches to human society. Solving perceived ecological problems undoubtedly demands a transformation of human attitudes and institutions. It may be argued that right-wing environmentalists have embraced all our points of definition without being 'Greens' in the fullest modern sense, given the left-wing nature of much contemporary Green politics. There have been a variety of forms of environmental politics, right, left and centre. 'Environmentalism', has been used as a term to describe a conservative philosophy that saw individuals and societies as strictly conditioned by environmental conditions, with little or no room for personal or social change outside set parameters of climate, soil and vegetation. The National Socialist doctrine of 'Blood and Soil' can be seen as an extreme but logical outcome of such 'environmentalism'. Many political ideologies, without endorsing a Green perspective, or even expressing a concern for environmental degradation, have used 'nature' to produce rules appropriate to human society. For example, although Marx emphasized the Promethean element of his doctrine in freeing humanity from the control of non-human and alien biological forces, his collaborator Engels attempted in *Anti-Dühring* to discover dialectical natural laws that illustrated the workings of all levels of reality, from human history to ecology, physics and chemistry. Hitler argued that he was obeying 'laws of nature' and so in a sense did Stalin (Gasman 1971). Social Darwinists have long argued that human societies evolve in a biological fashion, with the strongest reaching the top of natural and social communities by treading on the rest. Advocates of 'conservation' from Malthus to the US biologist Garrett Hardin have used scientific arguments to justify 'natural' social and economic inequality (Chapter Nine). In contrast Bookchin argues that radical environmentalists from Kropotkin onwards have rejected the conservative view of a vicious, mean, competitive nature. In his *Mutual Aid* the anarchist prince sought to show that neither society nor nature could survive without co-operative, as well as competitive, forms of behaviour (Chapter Eight). In contrast Grundman criticizes all such attempts to create a 'naturalistic' political philosophy:

> We find it in conservative authors like Gruhl; in Stalinist–Communist countries; and in eco-socialist writers like Lalonde. All claim the authority of nature and her laws to be the foundation stone of a new society that will solve ecological problems. Gruhl and Harich are alike in that they stress the iron necessity with which nature operates; from this they derive tough political measures. Bookchin argues that spontaneity in life converges with spontaneity in nature, and Lalonde stresses the fact that nature is, and society should be, self-organizing. . . . each version of nature . . . [is] a construction of its author.
>
> (Grundman 1991: 114–15)

Whether human institutions obey (or should obey) the same 'rules' as nature, and whether the natural sciences and social sciences are ultimately governed by the same principles, are interesting, much debated, questions beyond the scope of our attempt to describe briefly a 'Green' ideology. It is clear, though, that whether independently or via observing nature contemporary Greens adhere to a number of basic political and social principles. To 'ecology' the German and British Green parties added the fundamentals of 'ecology, social justice, non-violence and decentralisation of society' (Porritt 1984: 10–11), clearly differentiating their doctrine from those of both right-wing environmentalists and state socialists. Perhaps, despite participation in elections and parliaments, we may describe Green parties as anarchistic, or nearly so. Although critical of their love of technology and 'productivism', Greens equally embrace something of the so-called utopian socialist tradition of Owen and Fourier. Gandhi, with his non-violent strategy, integration of Hindu spirituality with politics, advocacy of grass-roots 'village socialism' and vegetarian respect for non-human life, may be described as a practitioner of an ecological politics. A Venn diagram composed of the overlapping circles of 'deep' ecological concern, socialism, and decentralism provides a territory within which we may find the Green ideology. Greens strongly oppose capitalism, as do many conservative environmentalists, and support social equality but are hostile to forms of socialism that are centralized and seek to dominate the natural world. Rather than being 'neither left nor right', the Greens by most forms of assessment are firmly on the left. Yet, clearly, the Green movement measures its philosophy with additional yardsticks: holism versus reductionism, animal liberation versus anthropocentrism, decentralism versus centralism, etc. Jonathan Porritt, a former leading member of the UK Green Party, provides a useful table of such distinguishing features of 'a Green paradigm' (1984: 216–17). He somewhat overstates his case by proclaiming that such a politics 'embraces every dimension of human experience and all life on Earth – that is to say, it goes a great deal further in terms of political comprehensiveness than any other political persuasion or ideology has ever

Distinguishing features of a 'Green paradigm'

The politics of industrialism	*The politics of ecology*
A deterministic view of the future	Flexibility and an emphasis on personal autonomy
An ethos of aggressive individualism	A co-operatively based, communitarian society
Materialism, pure and simple	A move towards spiritual, non-material values
Divisive, reductionist analysis	Holistic synthesis and integration
Anthropocentrism	Biocentrism
Rationality and packaged knowledge	Intuition and understanding
Outer-directed motivation	Inner-directed motivation and personal growth
Patriarchal values	Post-patriarchal, feminist values
Institutionalized violence	Non-violence
Economic growth and GNP	Sustainability and quality of life
Production for exchange and profit	Production for use
High income differentials	Low income differentials
A 'free-market' economy	Local production for local need
Ever-expanding world trade	Self-reliance
Demand stimulation	Voluntary simplicity
Employment as a means to an end	Work as an end in itself
Capital-intensive production	Labour-intensive production
Unquestioning acceptance of the technological fix	Discriminating use and development of science and technology
Centralization, economies of scale	Decentralization, human scale
Hierarchical structure	Non-hierarchical structure
Dependence upon experts	Participative involvement
Representative democracy	Direct democracy
Emphasis on law and order	Libertarianism
Sovereignty of nation state	Internationalism and global solidarity
Domination over nature	Harmony with nature
Environmentalism	Ecology
Environment managed as a resource	Resources regarded as strictly finite
Nuclear power	Renewable sources of energy
High energy, high consumption	Low energy, low consumption

Source: Porritt (1984: 216–17).

gone before'. None the less his survey does illustrate how a politically engaged Green defines his own philosophy, providing us with one set of distinguishing features with which to explore the existence or otherwise of an early Green politics and philosophy.

Using Porritt's criteria, we can detect a variety of historical movements Green in all but name. A well researched historical example of the Green package can be found in Gould's study of late nineteenth-century British radicalism, where he argues that 'the most fecund and important period of green politics before 1980 lay between 1880 and 1900' (1988: viii). During this period the designer, novelist and poet William Morris, drawing upon the Romantic conservatism of Carlyle and Ruskin, an English tradition of utopian socialist experiments such as the Chartist Land Company, as well as Marxism and anarchism, integrated Green themes with leftism. Despite the disdain of 'scientific socialists' such as Engels and H. M. Hyndman, who joined him in Britain's first Marxist party, the Social Democratic Federation, Morris clearly espoused a radical Green politics. Together with others, including the mystic and early advocate of gay rights Edward Carpenter, the animal liberationists Henry and Kate Salt, as well as the populist orator and publisher Blatchford, he vigorously proclaimed a philosophy virtually identical to Porritt's. Gould, comparing their views and those of today's Green parties, notes, 'They shared all but two of the twenty-nine features that Porritt identifies as distinguishing the politics of ecology and those two are the outcome of technological development' (1988: 161). *The Clarion*, a populist socialist paper, could, for example, proclaim in its issue of 14 April 1894 that 'To make the slum child love and long for nature is to sow the seed of revolution.' Carpenter, who practised near self-sufficiency with his lover George Merriel and founded the Sheffield Socialist Society, wrote that 'The vast majority of mankind must live in direct contact with nature.' Morris was hostile to urbanism and pollution, produced an early ecotopian novel, *News from Nowhere*, that looked to the creation of a pastoral England. Carpenter opposed a reductionist scientific method, endorsing 'intuition' and a holistic philosophy. Gould's 'back to nature' activists set up communes at Purleigh in Essex and Whiteway in Gloucester. Morris wished to link politics with a demand for the abolition of boring, soul-destroying labour and its replacement with creative work. One might continue almost *ad nauseum* to list clearly Green features of the late nineteenth-century eco-socialists. In 1884 the dispute between these 'utopians' and 'scientific socialists' created a split in the Social Democratic Federation, with Hyndman complaining that the socialist movement was in danger of becoming a 'depository of old cranks: humanitarians, vegetarians, anti-vivisectionists, arty-crafties and all the rest of them'. Although the Morris style of ecological socialism was defeated by a 'scientific', centralized and industrial doctrine, his views reflected the concern of large numbers of individuals both within the labour movement and in wider society. Between 1880 and 1900 Gould lists the creation of numerous conservation and environmental groups, including 'the Edinburgh Environment Society (1884), Society for Checking the Abuses of Public Advertising (1873), and the Coal Smoke Abatement Society (1898)' (1988: 16). Members of all strata of late Victorian society were intensely interested in environmental and land issues. Contemporary Greens might do well to investigate the reasons for both the rapid initial growth and the subsequent disappearance of the radical environmental politics of Morris and Carpenter.

Other Green movements have existed in the past. We know, for example, that in 1524 the rebellious Anabaptist Thomas Munzer, while swapping polemics with Luther, argued that it was intolerable that 'all creatures have been made into property, the fish in the water, the birds in the air, the plants on the earth – all living things must also become free' (Marx 1977: 239). Such sentiments were echoed centuries later by the New England poet Henry David Thoreau, who, although rather more isolated than Morris or Carpenter from wider society, clearly adhered to the Green package. Thoreau is the most interesting, because he combined social awareness with environmental concern, implacably opposed as he was to slavery, committed to civil disobedience in the pursuit of justice and familiar with Hindu and Buddhist ideas on nature. In 1861 he observed, with a note of sardonic prophecy, 'Thank God men cannot fly yet, and lay waste the sky as well as the Earth. We are safe on that side for the moment' (quoted in Pontin 1971: 15). The Pre-Raphaelites had some Green sensibilities, and Ruskin, although essentially rather conservative, was a formative influence on Morris. The left-wing English Romantics of the late eighteenth century, especially Blake and Shelley, Mary Shelley and in his younger years Wordsworth, were rather stronger and more politicized in their Green outlook than the Pre-Raphaelites. Shelley, for example, defended vegetarianism not just as an animal liberationist but equally using the rather modern-sounding ecological argument that more food might be produced per acre from plant crops than from animal husbandry (Chapter Ten). Gandhi claimed to have been inspired in his campaigns of resistance in South Africa and India by the poet's advocacy of non-violent direct action. Malthus argued in his famous essay on population that an increasing number of human mouths would, like so many locusts, lay the Earth bare. Shelley condemned him for propounding such a view, because, while the poet agreed with the concept of limiting growth for the sake of nature (he was vegetarian for just such a reason), he feared that Malthus would punish the poor for the mistakes of those with wealth, 'that the rich are still to glut . . . and that the poor are to pay with their blood, their labour, their happiness, and their innocence, for the crimes and mistakes which the hereditary monopolists of the earth commit' (Shelley 1965: 266). Redistribution and restrictions upon those who over-consumed at the expense of the poor and the planet, rather than sanctions against those with least, were his preference, and remain those of the modern Green movement when faced by conservative latter-day Malthusians. In short, Shelley's support for a form of grass-roots anarcho-socialism cannot be doubted. Pantheism, drawn from a Neoplatonist perspective, further adds to his radical ecological credentials. Mary Shelley wrote one of the first warnings against scientific knowledge and technology out of control in her Gothic novel *Frankenstein*. In turn her mother, Mary Wollstonecraft, was an early advocate of women's liberation, while her father, William Godwin, may, despite his rather uncritical adherence to science, be seen as a prime source of Green anarchism. Blake was a vital and clear advocate of a holistic philosophy and the sworn enemy of the reductionism of the Green *bêtes noires* Descartes and Newton. Edward Carpenter embraced, in the novelist E. M. Forster's poignant phrase, 'the socialism of Shelley and Blake' (1965: 217). Perhaps such a socialism captures the essence of all radical Green politics.

Other movements sharing most, if not all, of Porritt's criteria, may be identified. Reaching back as far as it is possible to go with our species, prehistory provides us with the shamanic myth of the original ecologist in the form of the hunter–gatherer. Although the supposed environmental credentials of native peoples have often been romanticized and are the subject of some critical debate, there is at least some truth in the claim. The Ainu, Innuit, Kalahari bushpeople, native Americans and similar groups adhere to a holistic outlook, practise forms of communistic direct democracy and respect nature. A wide variety of spiritual traditions may be seen as 'Green', including Taoism, Sufism, Zen and (more doubtfully) the more traditional forms of Buddhism. In much Eastern mysticism we again find holism, respect for other species and occasionally a link with political movements. In his survey of the Taoists, Needham notes traits of anarchism, pacifism and reverence for nature, essential to the contemporary Green movement in ancient China (1956, 2: 98). Today, in countries such as Thailand and Japan, many Buddhist monks are active in Green groups. Starhawk makes a spirited case for witchcraft as a link with the worship of the Earth Mother goddess, the original Green religion (Chapter Fifteen). A case may be made for describing a great many other creeds and political movements as 'Green'.

Individual elements of the Green canon have a past, separate from such possibly early Green or quasi-Green movements. In understanding the evolution of Green ideas we may be better placed to assess their importance (or otherwise) as elements within present-day radical environmentalism. A rural politics, defending both peasant and land, can be traced in Britain from Belloc and Chesterton, early twentieth-century Catholic advocates of the campaign for three acres and a cow, back to the Chartist land agitators and William Cobbett. Beyond Cobbett we reach the great tradition of peasant revolutionaries encompassing John Ball in the fourteenth century and Winstanley's Diggers in the seventeenth. Parallels of grass-roots peasant populism abound the world over, from ancient China to 1930s California. None of these ruralist movements was 'Green' but all, it may be argued, contributed to later Green movements. Holism, essential to the Green outlook, was developed by individuals often hostile to other elements of radical environmentalism such as Hegel and Marx. As Worster notes, even ecology in its long development as a science did not always act as a friend of the Earth. While it is possible, and necessary, to construct elaborate intellectual family trees, indicating the evolution and interconnection of discrete 'Green' themes, in examining Green history we need to go further than mere taxonomy. It is important to ask why these ideas waxed and waned.

This volume should be seen as a very preliminary introduction to the history of Green movements and Green ideas. Although an increasing number of books have appeared on the subject of environmental history, studies of specific groups and individuals adhering to a philosophy of radical, engaged ecological action are still rare. Equally, there has been relatively little research into the origins of contemporary Green movements. Here I have tried to provide space for early Green voices to speak their forgotten message. I have covered a series of themes, seen as essential to the modern movements, including deep ecology/animal liberation, scientific ecology, non-violent protest, holism, sustainable economic development and the Green critique of economic growth. I have included individuals, who I feel can clearly be labelled Green, such as Carpenter, Shelley and Morris, as well as others who adhered to elements of the philosophy. For example, I have included conservationists such as Evelyn and Pinchot, both of whom can be seen as civil service environmentalists who, although hostile to the fundamentals of radical deep ecology, sought solutions to environmental problems such as air pollution and deforestation.

The first chapter examines the interesting debate around the question of the existence, or otherwise, of the original prehistoric ecologist. It is a popular myth that humanity at one time lived in total ecological harmony (a difficult concept, in itself, to define) prior to some kind of Fall. I have included sources that both support and criticize belief in the Green credentials of our ancestors and of contemporary hunter–gather groups. Chapter Two briefly describes the ecological attitudes of early urban civilizations, especially those of Greece and Rome, that had such a formative influence on modern society. Chapter Three seeks to illustrate, together with some of the extracts from the previous sections, that environmental problems are very much older than many of us normally imagine. Chapter Four contains a number of attacks on the fundamentals of the Green ideology, from Agricola's defence of mining against detractors who feared for the Earth deity to the Webbs' enthusiasm for Stalin's remoulding of nature. Much of the rest of the volume deals with particular Green beliefs with a long pedigree. I have tried to include a taste of the different forms that Green philosophy has taken, including sections on feminism, literature, religion and revolutionary politics. Historically feminism has contained a strongly Green element and some feminists argue that the breakdown of matriarchical societies in prehistory led to the collapse of a harmonious ethic of balance between humanity and nature. Others have argued that new spiritual values are needed to restore the Earth or have looked to radical direct action.

Introduction

The final chapter, 'Utopia or bust', is inspired by the intuition that Green movements, far from attacking the modern and looking to some past perfection, perhaps in the age of the *Rainbow Warrior* or in an imagined medieval pastoral England, seek inspiration to overleap their age. The most radical Greens have always worked to construct a richly different form of society. Meaningful solutions to our severe environmental problems, the product not just of misjudged relations between humanity and nature but equally of bad blood between different sections of society, will come as much from literature and history as from science and economics. Dumont's manifesto for the 1974 French presidential election provides a comparatively recent piece of writing which is both practical and utopian. In summarizing so many elements of a radical ecological politics Dumont provides us with an appropriate concluding extract. The preceding chapter looks at the origins of Green party politics in the twentieth century. Finally I have included a guide to further reading that allows of greater exploration than is possible here.

The Green movement has deep and varied roots. The study of its history is a young discipline, hence an anthology such as this maps merely the thickest girths of a few roots; some plunge into the darkness and others have been mentioned only in passing. The very definition of Green politics and philosophy remains the subject of intense debate, yet, as we have seen, there can be no dispute that Green ideas are far older than is usually acknowledged. Many have struggled for the liberation of nature and of the human spirit, and it is hoped that their words contained in this survey will inspire not just study but new growth. Read, enjoy – there are some splendid pieces of writing included in this collection – and act!

Introductory extract
Alice Walker on the MOVE massacre

MOVE's work, John Africa's revolution, is to stop man's system from imposing on life. MOVE's work is to stop industry from poisoning the air, the water, the soil and to put an end to the enslavement of life – people, animals, *any* form of life.

(MOVE 1986)

It is Monday, 13 May 1985. At approximately 5.25 p.m., a Pennsylvania state police helicopter hovers sixty feet over 6221 Osage Avenue. Harnessed securely to the inside of the chopper's cabin, Lieutenant Frank Powell, commander of the Philadelphia Bomb Disposal Unit, the 'bomb squad', leans outside and hurls a green canvas bag towards the roof below. Extending from the bag is a lit 45-second fuse attached to a bomb.... Inside the house are thirteen MOVE people, seven adults, six children. On impact, the bomb throws off a fierce wave of heat of 7,200 fahrenheit, melting tar roof materials into a flammable liquid and turning wooden debris into flying kindle ... Glass windows half a block away completely shatter ... Of the thirteen people inside 6221 Osage, eleven are dead. Mangled, burned, carried away in zippered nylon bags – mostly in pieces.

(Harry 1987: 5–6)

Introduction

Alice Walker's essay 'Nobody was supposed to survive', examining the events surrounding a police assault on a Philadelphia house used by a small group of Afro-American utopian radicals, illustrates a number of salient themes in Green history. It reminds us that ideas never stand alone from individuals. Even if such individuals are 'idealists', refusing to let their children watch television, rescuing animals from laboratory experiments, piling up organic rubbish, to the disgust of their neighbours, their idealism is not drawn from thin air or the written word but is a product of their life experience. The Movement, more commonly known as MOVE, bombed out of their commune with FBI explosives, were a part of late twentieth-century America, their ideology and the enthusiasm with which they pursued it propelled by the failures they perceived in American society. While it is possible to trace intellectual fault lines and show how ideas have passed from generation to generation, all our early Greens or environmentalists need to be examined in a lived context if we are to understand how their beliefs took hold, evolved and were transmitted. For example, Morris gained much from Ruskin and the Romantic poets but was also strongly influenced by his experience of the industrialization of his day, which he felt was impoverishing the environment and exploiting the workers, when he became an eco-socialist in the 1880s. Environmental conditions – with MOVE, the social and ecological poverty of modern city life, including that of their own relatively prosperous neighbourhood – are an important factor influencing the growth of ecological politics.

The MOVE massacre also indicates how threatening radical Greens' ideas can be to the common sense of late capitalist society. Conservationists are seen as conservatives keen to keep grouse moors and salmon streams safe from the working class and secure for the exclusive use of an elite. While the competition for resources, whether consumer items, energy sources, land or water, can be used to define the problem of choice at the heart of economics and politics, Greens, often dismissed as apolitical or reactionary by the traditional left, extend such debates in a radical and challenging direction. Not merely content to ask for a fair share of worldly goods for a class or region, radical environmentalists demand that resources be conserved for future generations. Furthermore, in an obvious break with Marx and other left traditions, they confer ethical status upon the living beings – animals, plants and, more abstractly, nature – normally seen as resources available for unlimited exploitation by humanity. Radical Greens introduce some profoundly uncomfortable notions into the community, even when, unlike MOVE, they perfume their compost to maintain sweet relations with suburban neighbours. The exploitation of class by class, black by white and South by North in a global context, are seen as part of a process whereby an economic system exploits nature and destroys its ability to provide food, water, oxygen and the other necessaries of life. As Harry notes:

> Practically since its inception, MOVE has been attacked by various authorities and the mass media. Members have been described as filthy, their homes as vermin-infested, and they have been consistently portrayed as a violent back-to-nature cult led by a supposedly messianic madman known as John Africa . . . [because it] refuses to respect present-day America and its prevailing values. Its members openly defy official power and tirelessly preach against a system they consider utterly corrupt and destructive of life on this planet. . . .
>
> (Harry 1987: 7)

Greens struggle and have been killed in attempts to reverse such a process (Day 1989). Two years after the MOVE massacre, French special agents bombed the Greenpeace vessel *Rainbow Warrior*, with yet more loss of life. Anti-nuclear activists Hilda Murrell (UK, 1984) and Karen Silkwood (US, 1974) are examples, perhaps nearer to home, to environmentalists killed in extremely suspicious circumstances. Day argues that there is a global 'eco' war unfolding, with figures including Chico Mendes, the Brazilian labour activist, assassinated because of his efforts to protect the rain forests in 1988, at the pinnacle of a pyramid of fatalities. At its base are the millions, principally from the south of the globe, killed or injured by dumped toxic waste, landslides cause by deforestation, misused pesticides, poisoned water, air pollution and other ecological ills.

Finally, the MOVE massacre reminds us that, far from being a product simply of 'the despised "Northern White Empire"' (Bramwell 1989: 236), Green sensibilities are multiethnic and multi-cultural, with a wide appeal. As Alice Walker notes, the ideology of MOVE illustrates that 'poor people, not just upper- and middle-class whites and blacks who become hippies, are capable of intelligently perceiving and analysing American life, politically and socially ...' and reaching Green conclusions. Far from acting as victims, radical Greens in the South such as the Chipko movement in India, who hug trees to stop them being felled, are fighting back for planet and people with increasing success (Omvedt 1987: 29–38). In any anthology of this type there is an inevitable bias towards social groups and individuals who were successful, prosperous, literate and literary. It is easy to remember the poets and novelists whilst forgetting the peasants and native peoples who have been silenced or ridiculed out of the historical record. The work of intellectual and social archaeology required to give them voice and strengthen present communities of resistance has only just begun.

Alice Walker

'Nobody was supposed to survive'

'Nobody was supposed to survive.' – Ramona Africa
(*New York Times*, 7 January 1986)

I was in Paris in mid-May of 1985 when I heard the news about MOVE. My traveling companion read aloud the item in the newspaper that described the assault on a house on Osage Avenue in Philadelphia occupied by a group of 'radical, black, back-to-nature' revolutionaries that local authorities had been 'battling' for over a decade. As he read the article detailing the attack that led, eventually, to the actual bombing of the house (with military bombing material supplied to local police by the FBI) and the deaths of at least eleven people, many of them women, five of them children, our mutual feeling was of horror, followed immediately by anger and grief. Grief: that feeling of unassuageable sadness and rage that makes the heart feel naked to the elements, clawed by talons of ice. For, even knowing nothing of MOVE (short for Movement, which a revolution assumes) and little of the 'City of Brotherly Love', Philadelphia, we recognized the heartlessness of the crime, and realized that for the local authorities to go after eleven people, five of them children, with the kind of viciousness and force usually reserved for war, what they were trying to kill had to be more than the human beings involved; it had to be a spirit, an idea.

But what spirit? What idea?

There was only one adult survivor of the massacre: a young black woman named Ramona Africa. She suffered serious burns over much of her body (and would claim, later in court, as she sustained her own defense: 'I am guilty of nothing but hiding in the basement trying to protect myself and ... MOVE children'). The bombing of the MOVE house ignited a fire that roared through the black, middle-class neighborhood, totally destroying more than sixty houses and leaving 250 people homeless.

There we stood on a street corner in Paris, reading between the lines. It seems MOVE people never combed their hair, but wore it in long 'ropes' that people assumed were unclean. Since this is also how we wear out hair, we recognized this 'weird' style: dreadlocks. The style of the ancients: Ethiopians and Egyptians. Easily washed, quickly dried – a true wash-and-wear style for black people (and adventuresome whites) and painless, which is no doubt why MOVE people chose it for their children. And for themselves: 'Why suffer for cosmetic reasons?' they must have asked.

It appeared that the MOVE people were vegetarians and ate their food raw because they believed raw food healthier for the body and the soul. They believed in letting orange peels, banana peels, and other organic refuse 'cycle' back into the earth. Composting? They did not believe in embalming dead people or burying them in caskets. They thought they should be allowed to 'cycle' back to the earth, too. They loved dogs (their leader, John Africa, was called 'The Dog Man' because he cared for so many) and never killed animals of any kind, not even rats (which infuriated their neighbors), because they believed in the sanctity of all life.

Hmmm.

Further: They refused to send their children to school, fearing drugs and an indoctrination into the sickness of American life. They taught them to enjoy 'natural' games, in the belief that games based on such figures as Darth Vader caused 'distortions' in the personalities of the young that inhibited healthy, spontaneous expression. They exercised religiously, running miles every day with their dogs, rarely had sit-down dinners, ate out of big sacks of food whenever they were hungry, owned no furniture except a few pieces they'd found on the street, and refused to let their children wear diapers because of the belief that a free bottom is healthier. They abhorred the use of plastic. They enjoyed, apparently, the use of verbal profanity, which they claimed lost any degree of profanity when placed next to atomic or nuclear weapons of any sort, which they considered *really* profane. They hated the police, who they claimed harassed them relentlessly (a shoot-out with police in 1978 resulted in the death of one officer and the imprisonment of several MOVE people). They occasionally self-righteously and disruptively harangued their neighbors, using bullhorns. They taught anyone who would listen that the US political and social system is corrupt to the core – and tried to be, themselves, a different tribe within it. . . .

. . . the city officials and MOVE neighbors appeared to have one thing in common: a hatred of the way MOVE people chose to live. They didn't like the 'stench' of people who refused, because they believe chemicals cause cancer, to use deodorant; didn't like orange peels and watermelon rinds on the ground; didn't like all those 'naked' children running around with all that uncombed hair. They didn't appreciate the dogs and the rats. They thought the children should be in school and that the adults and children should eat cooked food; everybody should eat meat. They probably thought it low class that in order to make money MOVE people washed cars and shoveled snow. And appeared to enjoy it.

MOVE people were not middle class. Many of them were high-school dropouts. Many of them were mothers without husbands. Or young men who refused any inducement to 'fit in'. Yet they had the nerve to critique the system. To reject it and to set up, in place of its rules, guidelines for living that reflected their own beliefs.

The people of MOVE are proof that poor people, not just upper- and middle-class whites and blacks who become hippies, are capable of intelligently perceiving and analyzing American life, politically and socially, and of devising and attempting to follow a different – and, to them, better – way. But because they are poor and black, this is not acceptable behavior to middle-class whites and blacks who think all poor black people should be happy with jherri curls, mindless (and lying) TV shows, and Kentucky fried chicken.

This is not to condone the yelping of fifty to sixty dogs in the middle of the night, dogs MOVE people rescued from the streets (and probable subsequent torture in 'scientific' laboratories), fed, and permitted to sleep in their house. Nor to condone the bullhorn they used to air their neighbors' 'backwardness' or political transgressions, as apparently they had a bad habit of doing. From what I read, MOVE people were more fanatical than the average neighbors. I probably would not have been able to live next door to them for a day.

The question is: Did they deserve the harassment, abuse, and, finally, the vicious death other people's intolerance of their life style brought upon them? *Every bomb ever made falls on all of us.* And the answer is: No.

In *Living by the Word*, London: Women's Press (1988), 155–7, 159–60.

CHAPTER 1

Ancient wisdom

P. S. Martin, *The discovery of America*

M. Sahlins, *The original affluent society*

W. Pennington, *The elm decline*

J. G. Frazer, *The worship of trees*

Introduction

These are the greater complexities. The eco-system of Salisbury Plain and in the Nile Valley was simpler and easier to manage. The ecology was balanced, the demands on natural resources minimal. There was an awareness, an awe and respect for the environment, and in those prehistoric times there was no doubt a sufficiently satisfying account of the place of man in the universe. In that ancient stable framework, man had achieved the essence of those goals that are now being set for civilization today.

(Hawkins 1977: 264)

In the great debate between Hobbes and Rousseau the Green movement has tended towards the latter's faith in a state of primitive harmony and nobility rather than the former's belief in brutality and misery. Greens and fellow travellers have used existing hunter–gatherer groups and their ancient ancestors as an example of ecological good conduct. Social and economic lessons are often also drawn. Conservatively inclined Greens argue that the natural social order of modern peoples, studied by anthropologists, provides a critique of the social practices of unnatural urban civilization. For Goldsmith *et al.* (1988) questions concerning the status of women, child-rearing, diet, economics and war can be answered best by looking at so-called 'primitive' societies. Such peoples lived in a state of nature and were therefore part of a natural order in a social sense. Such an essentialist view that uses biological 'laws' to determine social practice has long been a staple of conservative political philosophies. From the left, Marx and Engels, whilst celebrating industrial progress and scientific advance, noting the work of the anthropologist Lewis Morgan, looked back to a state of pre-agricultural 'Primitive Communism' (Krader 1979: 153–71). Both eco-socialists and Green conservatives note that prehistoric and existing hunter–gather groups developed economies based on sharing rather than competitive exchange. A close knowledge of local eco-systems, many modern anthropologists have argued, allows such groups to live in a state of prosperity in the harshest of environments such as the Australian or Kalahari deserts. Finally eco-feminists have claimed that archaeological research reveals that such early societies were ecological, equal and matriarchal (Gimbutas 1991). Whether by matriarchal, Marxist or conservative constructions most members of the modern Green movement argue that so-called primitive peoples lived in balance with their environment.

A history of the origins of Green ideas can, it seems, be constructed solely from the retelling of attitudes towards this supposed ecological Eden. The Romantic poets, with their more than occasional ecological sensibilities, were morally encouraged by Rousseau's 'noble savage' at the turn of the nineteenth century. Ernest Thompson Seton's (1860–1946) experiences amongst the Cree Blackfoot inspired him to set up the Woodcraft Folk, in its original guise a leftist-oriented Green movement in 1920 Britain, later a youth movement that exists to this day (Prynn 1983). Horkheimer and Adorno sought to inject Marxism with transcendentalist nature reverence during the 1940s, in an ambitious and contradictory project, with reference to primitive mythology. Goldsmith, who founded *The Ecologist* magazine and helped create the first modern ecology party, People in the early 1970s, was directly influenced by his own travels in Africa and the work of Jean Leidloff, who had lived with Amazonian natives before publishing her book *The Continuum Concept* (1975).

Paradoxically, Green politics often ignores its historical roots whilst perpetuating a myth of primitive and prehistoric pre-existence. On occasion it ignores concrete research in favour of comforting romantic myth. It has been known to make startling mistakes. The famous Chief Seattle speech, reprinted by the former UK Ecology Party, Friends of the Earth International, and used by the organisers of Earth Day in 1992, for example, has proved to be a fabrication. The speech, supposedly addressed to the US President in 1854, and including the statement that 'The Earth is our mother ... I have seen a thousand rotting buffaloes on the prairies left by the white man, who shot them from a passing train', was in fact written as part of a film script by a Texan author in 1971 (Lichfield 1992). Our first extract, by the US geo-scientist P. S. Martin, argues that Seattle's ancestors, the Pleistocene inhabitants of North America, exterminated species of large mammals, including the mammoth and mastodon. Giant kill sites found since the publication of his original article show that buffalo and other large mammals were stampeded over cliffs, killing many more creatures than could possibly have been consumed. Some prehistoric peoples damaged their local eco-systems, and species were hunted to extinction, not just in the Americas but also in New Zealand, Madagascar and possibly Europe (Black 1970: 12).

Prehistory is a big place and cannot be judged crudely as an era either of imagined harmony or of generalized destruction. An attitude that all contemporary hunter–gatherers are 'primitive' and awaiting progress must, surely be rejected. The idea that such groups, together with our Neolithic, Mesolithic and Palaeolithic forebears, were purely and simply Green is equally naive and ill informed. Case studies, however, provide some evidence that hunter–gatherers lived in greater harmony with their environment than other societies (Clarke 1976; Diamond 1974). Our second extract argues the case for an 'original affluent society' based on observation of Kalahari bush people and Australian aborigines. The third looks at the deforestation caused in Britain and Ireland by the first Neolithic farmers. The fourth extract, from Frazer's (1854–1941) monumental study of mythology *The Golden Bough*, records instances of tree worship. Anthropology and environmental archaeology will no doubt furnish evidence of the ecological realities of hunter–gatherer life.

P. S. Martin

The discovery of America

At some time toward the end of the last ice age, big game hunters in Siberia approached the Arctic Circle, moved eastward across the Bering platform into Alaska, and threaded a narrow passage between the stagnant Cordilleran and Laurentian ice sheets. I propose that they spread southward explosively, briefly attaining a density sufficiently large to overkill much of their prey.

Overkill without kill sites. Pleistocene biologists wish to determine to within 1,000 years at most the time of the last occurrence of the dominant Late Pleistocene extinct mammals. If one recognizes certain hazards of 'push-button' radiocarbon dating,[1] especially dates on bone itself, it appears that the disappearance of native American mammoths, mastodons, ground sloths, horses, and camels coincided very closely with the first appearance of Stone Age hunters around 11,200 years ago.[2]

Not all investigators accept this circumstance as decisive or even as adequately established. No predator–prey model like Budyko's[3] on mammoth extinction has been developed to show how the American megafauna might have been removed by hunters.[4] Above all, prehistorians have been troubled by the following paradox.

In temperate parts of Eurasia, large numbers of Paleolithic artifacts have been found in many associations with bones of large mammals. Although the evidence associating Stone Age hunters and their prey is overwhelming, not much extinction occurred there. Only four late-glacial genera of large animals were lost, namely the mammoth (*Mammuthus*), woolly rhinoceros (*Coelodonta*), giant deer (*Megaloceros*), and musk-ox (*Ovibos*).

In contrast, the megafauna of the New World, very rarely found associated with human artifacts in kill or camp sites[5] was decimated. Of the thirty-one genera of large mammals[6] that disappeared in North America at the end of the last ice age, only the mammoth (*Mammuthus*) is found in unmistakable kill sites. The seven kill sites listed by Haynes[7] lack the wealth of cultural material, including art objects, associated with the Old World mammoth in eastern Europe and the Ukraine. It is not surprising that some investigators discount overkill as a major cause of the extinctions in America.

But if the new human predators found inexperienced prey, the scarcity of kill sites may be explained. A rapid rate of killing would wipe out the more vulnerable prey before there was time for the animals to learn defensive behavior, and thus the hunters would not have needed to plan elaborate cliff drives or to build clever traps. Extinction would have occurred before there was opportunity for the burial of much evidence by normal geological processes. Poor paleontological visibility would be inevitable. In these terms, the scarcity of kill sites on a land mass which suffered major megafaunal losses becomes a predictable condition of the special circumstances which distinguish a sudden invasion from more gradual prehistoric cultural changes *in situ*. Perhaps the only remarkable aspect of New World archaeology is that *any* kill sites have been found. . . .[8]

Unless one insists on believing that Paleolithic invaders lost enthusiasm for the hunt and rapidly became vegetarians by choice as they moved south from Beringia, or that they knew and practiced a sophisticated, sustained yield harvest of their prey, one would have no difficulty in predicting the swift extermination of the more conspicuous native American large mammals. I do not discount the possibility of disruptive side effects, perhaps caused by the introduction of dogs and the destruction of habitat by man-made fires. But a very large biomass, even the 2.3×10^8 metric tonnes of domestic animals now ranging the continent, could be overkilled within 1,000 years by a human population never exceeding 10^6. We need only assume that a relatively innocent prey was suddenly exposed to a new and thoroughly superior predator, a hunter who preferred killing and persisted in killing animals as long as they were available.[9]

With the extinction of all but the smaller, solitary, and cryptic species, such as most cervids, it seems likely that a more normal predator–prey relationship would be established. Major cultural changes would begin. Not until the prey populations were extinct would the hunters be forced, by necessity, to learn more botany. Not until then would they need to readapt to the distribution of biomes in America in the manner Fitting[10] has proposed.

An explosive model will account for the scarcity of extinct animals associated with Paleo-Indian artifacts in obvious kill sites. The big game hunters achieved high population density only during those few years when their prey was abundant. Elaborate drives or traps were unnecessary.

Sudden overkill may explain the absence of cave paintings of extinct animals in the New World and the lack of ivory carvings such as those found in the mammoth hunter camps of the Don Basin. The big game was wiped out before there was an opportunity to portray the extinct species.

Notes

1   P. S. Martin, in *Pleistocene Extinctions: the Search for a Cause*, P. S. Martin and H. E. Wright, Jr., eds. (Yale University Press, New Haven, Conn., 1967), pp. 87–9.
2   Over the past two decades radiocarbon dates have been published which, if taken at face value, appear to show that mammoths, mastodons, ground sloths, and other common members of the extinct American megafauna lasted into the postglacial ... For a complete description of field and laboratory treatment of the samples and for laboratory designations, see *Radiocarbon* 9–13 (1967–1971).
3   M. I. Budyko, *Sov. Geogr. Rev. Transl.* 8 (No. 10), 783 (1967).
4   According to R. F. Flint (*Glacial and Quaternary Geology*, Wiley, New York, 1971, p. 778), 'The argument most frequently advanced against the hypothesis of human agency is that in no territory was man sufficiently numerous to destroy the large numbers of animals that became extinct.'
5   Apart from postglacial records of extinct species of *Bison*, very few kill sites have been discovered. J. J. Hester (in *Pleistocene Extinctions: the Search for a Cause*, P. S. Martin and H. E. Wright, Jr., eds., Yale University Press, New Haven, Conn., 1967, p. 169), A. J. Jellinek (*ibid.*, p. 193), and G. S. Krantz (*Amer. Sci.* 58, 164, 1970) have all raised this point as a counterargument to overkill.
6   The North American megafauna that I believe disappeared at the time of the hunters includes the following general: *Nothrotherium, Megalonyx, Eremotherium,* and *Paramylodon* (ground sloths); *Brachyostracon* and *Boreostracon* (glyptodonts); *Castorides* (giant beaver); *Hydrochoerus* and *Neochoerus* (extinct capybaras); *Arctodus* and *Tremarctos* (bears); *Smilodon* and *Dinobastis*

(saber-tooth cats); *Mammut* (mastodon); *Mammuthus* (mammoth); *Equus* (horse); *Tapirus* (tapir); *Platygonus* and *Mylohyus* (peccaries); *Camelops* and *Tanupolama* (camelids); *Cervalces* and *Sangamona* (cervids); *Capromeryx* and *Tetrameryx* (extinct pronghorns); *Bos* and *Saiga* (Asian antelope); and *Bootherlum, Symbos, Euceratherium,* and *Preptoceras* (bovids).

7   C. V. Haynes, in *Pleistocene and Recent Environments of the Central Great Plains,* W. Dort, Jr., and J. K. Jones, Jr., eds. (University of Kansas Department of Geology Special Publication No. 3, Lawrence, 1971), p. 77.

8   A. Dreimanis, *Ohio J. Sci.* 68, 257, (1968).

9   Even when most of their calories come from plants (see R. B. Lee, in *Man the Hunter,* R. B. Lee and I. Devore, eds., Aldine, Chicago, 1969), men of modern nonagricultural tribes devote much time to the hunt. The arctic invaders of America had come through a region notably deficient in edible plants. As long as large mammals were flourishing, there was no need to devise new techniques of harvesting, storing, and preparing less familiar food. None of their artifacts suggests that the first American hunters also stalked the wild herbs, and none of their midden refuse suggests that the succeeding gatherers knew the extinct mammals.

10   J. E. Fitting, *Amer. Antiquity* 33, 441. (1968).

'The discovery of America', *Science* 179 (1973), 969, 972.

Marshall Sahlins

The original affluent society

If economics is the dismal science, the study of hunting and gathering economies must be its most advanced branch. Almost universally committed to the proposition that life was hard in the paleolithic, our textbooks compete to convey a sense of impending doom, leaving one to wonder not only how hunters managed to live, but whether, after all, this was living? The specter of starvation stalks the stalker through these pages. His technical incompetence is said to enjoin continuous work just to survive, affording him neither respite nor surplus, hence not even the 'leisure' to 'build culture'. Even so, for all his efforts, the hunter pulls the lowest grades in thermodynamics – less energy/capita/year than any other mode of production. And in treatises on economic development he is condemned to play the role of bad example; the so-called 'subsistence economy'.

The traditional wisdom is always refractory. One is forced to oppose it polemically, to phrase the necessary revisions dialectically: in fact, this was, when you come to examine it, the original affluent society. Paradoxical, that phrasing leads to another useful and unexpected conclusion. By the common understanding, an affluent society is one in which all the people's material wants are easily satisfied. To assert that the hunters are affluent is to deny then that the human condition is an ordained tragedy, with man the prisoner at hard labor of a perpetual disparity between his unlimited wants and his insufficient means.

For there are two possible courses to affluence. Wants may be 'easily satisfied' either by producing much or desiring little. The familiar conception, the Galbraithean way, makes assumptions peculiarly appropriate to market economies: that man's wants are great, not to say infinite, whereas his means are limited, although improvable: thus, the gap between means and ends can be narrowed by industrial productivity, at least to the point that 'urgent goods' become plentiful. But there is also a Zen road to affluence, departing from premises somewhat different from our own: that human material wants are finite and few, and technical means unchanging but on the whole adequate. Adopting the Zen strategy, a people can enjoy an unparalleled material plenty – with a low standard of living.

That, I think, describes the hunters. And it helps explain some of their more curious economic behavior: their 'prodigality', for example – the inclination to consume at once all stocks on hand, as if they had it made. Free from market obsessions of scarcity, hunters' economic propensities may be more consistently predicated on abundance than our own. Destutt de Tracy, 'fish-blooded bourgeois doctrinaire' though he might have been, at least compelled Marx's agreement on the observation that 'in poor nations the people are comfortable', whereas in rich nations 'they are generally poor'.

This is not to deny that a preagricultural economy operates under serious constraints, but only to insist, on the evidence from modern hunters and gatherers, that a successful accommodation is usually made. After taking up the evidence, I shall return in the end to the real difficulties of a hunting–gathering economy, none of which are correctly specified in current formulas of paleolithic poverty. . . .

Another specifically anthropological source of paleolithic discontent develops in the field itself, from the context of European observation of existing hunters and gatherers, such as the native Australians, the Bushmen, the Ona or the Yahgan. This ethnographic context tends to distort our understanding of the hunting–gathering economy in two ways.

First, it provides singular opportunities for naïveté. The remote and exotic environments that have become the cultural theater of modern hunters have an effect on Europeans most unfavorable to the latter's assessment of the former's plight. Marginal as the Australian or Kalahari desert is to agriculture, or to everyday European experience, it is a source of wonder to the untutored observer 'how anybody could live in a place like this'. The inference that the natives manage only to eke out a bare existence is apt to be reinforced by their marvelously varied diets (cf. Herskovits, 1958, quoted above). Ordinarily including objects deemed repulsive and inedible by Europeans, the local cuisine lends itself to the supposition that the people are starving to death. Such a conclusion, of course, is more likely met in earlier than in later accounts, and in the journals of explorers or missionaries than in the monographs of anthropologists; but precisely because the explorers' reports are older and closer to the aboriginal condition, one reserves for them a certain respect.

Such respect obviously has to be accorded with discretion. Greater attention should be paid a man such as Sir George Grey (1841), whose expeditions in the 1830s included some of the poorer districts of western Australia, but whose unusually close attention to the local people obliged him to debunk his colleagues' communications on just this point of economic desperation. It is a mistake very commonly made, Grey wrote, to suppose that the native Australians 'have small means of subsistence, or are at times greatly pressed for want of food'. Many and 'almost ludicrous' are the errors travellers have fallen into in this regard: 'They lament in their journals that the unfortunate Aborigines should be reduced by famine to the miserable necessity of subsisting on certain sorts of food, which they have found near their huts; whereas, in many instances, the articles thus quoted by them are those which the natives most prize, and are really neither deficient in flavour nor nutritious qualities.' To render palpable 'the ignorance that has prevailed with regard to the habits and customs of this people when in their wild state', Grey provides one remarkable example, a citation from his fellow explorer, Captain Sturt, who, upon encountering a group of Aboriginals engaged in gathering large quantities of mimosa gum, deduced that the '"unfortunate creatures were reduced to the last extremity, and, being unable to procure any other nourishment, had been obliged to collect this mucilaginous."' But, Sir George observes, the gum in question is a favorite article of food in the area, and when in season it affords the opportunity for large numbers of people to assemble and camp together, which otherwise they are unable to do.

Stone Age Economics, Chicago: Aldine Atherton (1972), 1–2, 6–7.

W. Pennington

The elm decline

In 1941, while the elm decline was still generally attributed to climatic change, and before the power of prehistoric man to dominate his environment had been realised, Iversen published a classic paper attributing certain vegetation changes found just *above* the elm decline to deliberate forest clearance by Neolithic agriculturalists. The pollen curves suggested a clearance of all trees in a limited area, by felling and burning (revealed by a charcoal layer), followed by a primitive form of cultivation of cereals in the cleared patches. Regeneration of the forest followed quite quickly, with colonisation by pioneer trees succeeded by re-establishment of the high forest, in its original form. Iversen attributed this vegetation succession to shifting cultivation by nomadic farmers who moved on to fell a new patch when the first fertility of the cleared ground was exhausted. This, which Iversen called a 'Landnam' clearance, differed from the elm decline in that at the clearance level *all* the forest trees were affected – that is, the absolute quantity of tree pollen relative to that of other plants was very much diminished. Iversen interpreted this as wholesale clearance of a patch of mixed forest. The pollens which showed an increase in percentage of the total were grasses, and weeds such as *Plantago lanceolata*, *Rumex*, *Artemisia*, and members of the Chenopodiaceae, and then bracken. At the same time, pollen of the cultivated cereals appeared for the first time. Then regeneration of the forest proceeded quite quickly, with colonisation first by birch and other pioneer trees, which were succeeded by oak and lime, and some elm – the trees of the high forest. No selective exploitation of any one tree is involved, and the succession of pollens gives a very vivid picture of the felling, the primitive cultivation, and the invasion of weeds and bracken, and then the stages of return to the original forest. This is a typical Landnam.

At that time (1941) many workers were reluctant to admit that the necessarily small numbers of prehistoric people could have such a profound effect on the local pollen rain, and efforts were made by some to interpret these changes in Sub-boreal pollen curves as due to some climatic change of small amplitude. In 1953, the Copenhagen pollen analysts carried out a field experiment to demonstrate that it was possible for a small group of men, equipped only with Neolithic axes, to cut down mature trees and clear by burning a fair-sized patch of established forest, within about a week. Polished stone axes from the Copenhagen National Museum were fitted with new wooden hafts, made on the model of those which have been preserved in peat bogs from Neolithic times, and it was found that, using these, three men could clear about 600 square yards of forest in four hours. Illustrations of this field experiment, which was carried out at Draved Forest in South Jutland, can be found in Professor Ovington's book on Woodlands in this series. There will be found also photographs of the crop of *Triticum dicoccum* (the Emmer wheat which is known to have been cultivated throughout the prehistoric period) which was raised in the woodland soil plus the ashes of the burning, in the year of the clearance (1953), contrasted with the poor crop of the same plant obtained three years after the clearing and burning, when declining soil fertility clearly made it no longer profitable to try to grow cereals there. This doubtless explains the apparent short duration of the Landnam episodes.

This experiment, and the widespread discovery of similar Landnam episodes in many places where the necessarily very closely spaced sampling was carried out, convinced botanists that the Neolithic agriculturalists (not necessarily, indeed almost certainly not, the same cultures as those responsible for the elm decline) were able to destroy, at least temporarily, very considerable areas of primary forest. The polished stone axes of the type used in the Draved experiment were made of flint, and this was apparently a favoured material when available. In those parts of Britain where no flint was available, certain types of igneous rock were found to have the same hardness and response to flaking and grinding. The distribution of these axes, as shown by recorded finds, reveals that there must have been a very considerable trade in them, and suggests the extensiveness of forest destruction which must have resulted from their use . . .

When British pollen diagrams for the early part of the Sub-boreal (Zone VIIb) are considered as a whole, it seems likely that the effects of the successive Neolithic cultures . . . must have altered the composition of the primary forest over most, if not all, of the British Isles, with the possible exception of parts of Scotland, where there is not enough data, as yet, to be sure. The initial disaster to the elm at the elm decline was followed, in some areas, by more or less regeneration of this tree to something approaching its former position as a contributor to the total tree pollen. On poorer soils, likely to have been already impoverished by leaching, elm never regained its status, and presumably was never again an important tree in the regenerated forests. On calcareous soils, as in central Ireland and on the limestone round Morecambe Bay, the percentage of elm pollen was restored after Neolithic forest clearances to something near its former level.

Another part of Britain where Neolithic clearance led to what appears to have been more or less permanent deforestation was on the coastal plain of south-west Cumberland, a fertile strip of drift-covered lowland between the Lakeland mountains and the sea. In the kettle-holed surface of the drift are many small depressions, some of which contain tarns, and others are now fens or *Sphagnum* bogs. Ehenside Tarn which is a rather shallow tarn which was partially drained in the nineteenth century, is one of the few settlement sites of Neolithic age known in north-west England. The draining of the tarn revealed hearths, pottery, artefacts of wood and bone, a saucer quern, and polished and unpolished stone axes which have comparatively recently been petrologically identified as of Great Langdale origin and Cumbrian type. The outcrops of a particular hard tuff in the Borrowdale volcanic rocks, high on Langdale Pikes and on Scafell and Scafell Pike, were extensively worked to produce these tools. No complete polished axes have been found at the factory sites, but only rough-outs and flakes, so it has been supposed that the rough-outs were finished at lowland settlement sites such as Ehenside Tarn, where sandstone grinding slabs and rubbers were found. Pollen diagrams from the deep muds of Ehenside Tarn, and from three other sites on the coastal strip – Mockerkin Tarn, north of Ehenside, Barfield Tarn, at the southern extremity, and on the sand-dunes at Eskmeals – all agree in indicating a reduction in tree pollen, beginning at the Elm Decline, which seems to indicate a progressive and permanent change in vegetation, from oak woods with elm and birch, with alder and willow in the wet hollows, to progressively increasing grassland with considerable expansion of bracken.

The History of British Vegetation, London: English Universities Press (1969), 60–73.

Sir James Frazer

The worship of trees

The worship of trees is very widespread among the tribes of French or West Sudan. Thus, for example, among the Bobo, at the time of sowing, the chief of the village offers sacrifices in the field to any large trees that happen to be there. Each of these trees represents at once the Earth and the Forest, two great and powerful divinities which in the mind of the negro form a single great divinity. Thus the sacrifice is offered at the same time to the Earth and the Forest that they may be favourable to the sowing. The victim sacrificed on these occasions is a hen, or several hens. But these are not the only sacrifices offered by the Bobos to trees. At Kabourou there is a chief who owns a sacred tree, a wild fig. He alone has the right of making sacrifices to it. If another person wishes to make a sacrifice to it he can only do so by leave obtained from the chief. At Tone there are five large sacred trees in the village itself. After the harvest the chiefs of the village offer sacrifices to these trees, representing the Earth and the Forest, as a thank-offering for giving much millet, and for having warded off diseases, and so on. They offer hens to them. But contrary to the custom of many other tribes, the Bobos have no sacred woods or thickets.[1] Among the Menkieras, another tribe of the French Sudan, some people, but not all, offer sacrifices to trees. At Zinou and Bono the trees to which sacrifices are made are the *soun-soun, cailcedrats, karites,* and tamarinds, in which the spirit of the Forest is believed to reside.[2] On the day after a copious shower of rain has fallen among the Nounoumas, another tribe of the French Sudan, the chief of the village takes a hen into his field. If there is in the field a tamarind, a *karite,* or a *cailcedrat,* he pours the blood of the fowl over the tree. But if there is no such tree he pours the blood of the fowl upon the earth. The sacrifice is offered to the Earth and the Forest to procure a good crop. They also invoke the Good God or Heaven. In the mind of the negro the tree is, firstly, a child of the earth, since the Earth bears it upon its breast, and secondly a representative of the Forest, since the forest is formed of grass, plants, and trees. Thus to offer a sacrifice to a tree is at the same time to offer it to the Earth and the Forest, the two great divinities of productivity. That is why, when there is a tree in the field, they pour the blood of the hen upon it. Three divinities are thus honoured by this sacrifice: the inferior divinity of the tree itself, and the more powerful divinities of the Earth and the Forest. It is needless to explain why at time of sowing sacrifices are offered to the two latter divinities: it is the Earth which controls the growth of the grain, it is the Forest which is the divinity of vegetation in general.[3] The Kassounas-Fras, another tribe of the French Sudan, also offer sacrifices to trees. They have sacred trees, either at the gate of the village, or in the fields. Each of these sacred trees has its master, whose leave must be obtained by any person who wishes to sacrifice to the trees. They offer sacrifices every time that the diviner bids them do so. The sacrifice consists of a hen, millet-meal, and sometimes a small pebble. The Kassounas-Fras have also sacred woods. It is the Chief of the Earth, assisted by the elders of the village, who offers sacrifices to the sacred wood each time that the diviner tells him to do so. The sacrifice offered to the Sacred Wood is also offered to the Earth, of which the Sacred Wood is the child. In these sacred woods there are small heaps of stones at which these sacrifices are offered. At present one may walk in the sacred woods. Formerly it was severely prohibited to do so. But on no account may one cut wood there.[4]

A French officer engaged in surveying the country inhabited by the primitive Moïs of Indo-China witnessed one such ceremony of propitiation performed by the natives before felling a tree. He says, 'It sometimes happened in the course of our geodetical survey that we were compelled to cut down a tree which interrupted the field of view of our instruments. A most interesting scene preceded the act of destruction. The "foreman" of our Moi coolies approached the condemned tree and addressed it much as follows: "Spirit who hast made thy home in this tree, we worship thee and are come to claim thy mercy. The white mandarin, our relentless master, whose commands we cannot but obey, has bidden us to cut down thy habitation, a task which fills us with sadness, and which we only carry out with regret. I adjure thee to depart at once from this place and seek a new dwelling place elsewhere, and I pray thee to forget the wrong we do thee, for we are not our own masters."'[5]

Notes

1   L. Tauxier, *Le Noir du Soudan* (Paris, 1912), pp. 70 *sq.*
2   L. Tauxier, *op. cit.*, pp. 104 *sq.*
3   L. Tauxier, *op. cit.*, pp. 190 *sq.*
4   L. Tauxier, *op. cit.*, p. 237.
5   H. Baudesson, *Indo-China and its Primitive People*, p. 129.

Aftermath: a Supplement to 'The Golden Bough', London: Macmillan (1936), 126–7, 136–7.

CHAPTER 2

Ecology and early urban civilization

J. D. Hughes, Ecology in ancient Mesopotamia
Plato, Eroded Attica
Lucretius, The exhausted Earth
J. D. Hughes, Ecology in imperial Rome
J. Nriagu, Lead and lead poisoning in antiquity

Introduction

We are the absolute masters of what the earth produces. We enjoy the mountains and the plains. The rivers are ours; we sow the seed and plant the trees.
We fertilize the earth ... We stop, direct, and turn the rivers. In short, by our
hands we endeavour, by our various operations on this world, to make, as it
were, another nature.

(Cicero [106–43 B.C.], quoted by Hughes, 1975a)

There is no doubt that the earliest urban civilizations created severe environmental
damage. The accumulation of the agricultural surpluses necessary to build cities, construct
pyramids and temples, equip armies, provide regalia for priestly rulers and maintain
bureaucracies led to over-farming, soil erosion and the destruction of natural habitats.
Irrigation, in ancient Mesopotamia and the Indus valley, led to over-salinization as the
evaporation of large bodies of water left toxic salt on formerly fertile land in 'a Satanic
mockery of snow' (Goudie 1981: 113). Huge areas of forest were removed to make way for
fields of wheat. Loss of tree cover and over-enthusiastic farming practices resulted in
widespread soil erosion.

Hughes (1975a) examines the problem of salinization and looks in detail at the environmental ill effects of early city life. In Mesopotamia water was likely to be polluted, smoke
from thousands of fires poisoned the air and the death rate was presumably high.
Classical Greek society, thousands of years later, was equally afflicted. In our second
extract we find the philosopher Plato (428–347 B.C.) bemoaning the damage done to the
land by soil erosion.

The attitudes of early societies, especially those of classical Greece, strongly influence
contemporary approaches to the environment. The Mesopotamians saw nature as a wild
beast to be tamed. Aristotle (384–322 B.C.) and the early Stoics claimed that nature was a
resource placed before humanity for its exclusive use. Already at this early date anthro-
procentric arrogance seemed to be the dominant trend. In *The White Goddess* Graves argues
(Chapter Sixteen) that the rationalism of such philosophers had swept away an earlier
pantheistic appreciation of the Earth. Some Greeks continued to worship Pan and Gaia.
Others agreed that nature, even if it was to be used primarily for human gain, should at
least be conserved and maintained with wisdom. A Greek conservationist movement is
tentatively identified by Hughes, who points to environmentally conscious critics of
Aristotle and practices of protecting woodland by declaring areas of forest as sacred
groves. Green philosophical holism, as we shall see, may in part be traced back to the

ideas of Heraclitus. Pythagorus (c. 580–c. 500 B.C.) was both a mathematician and a pioneer of vegetarian mysticism (see Chapter Five). Sappho (610–580 B.C.) and other poets wrote odes to nature, while of course the concept of Gaia the Earth mother may be traced to an ancient Greek tradition. Obviously it is possible here only to hint at the complexity of Greek attitudes to the environment.

Rome, especially in its days of empire, had fewer defenders of the Earth. In our next extract we find Titus Lucretius Carus (c. 99–55 B.C.), author of a study of nature, musing that the Earth is in a state of decay and is no longer fruitful. The ill effects of Rome's exploitative agriculture may lead to symptoms of diminished fertility. Again Hughes provides an exhaustive account of Roman insults to the natural world. As well as the usual soil erosion, deforestation and pollution, the citizens of Rome unleashed an astonishingly brutal assault on the animal kingdom, with gladiatorial contests and imperial celebrations leading to the extermination of fantastic numbers of beasts. 'At the dedication of the Colosseum under Titus, 9,000 were destroyed in 100 days, and Trajan's conquest over Dacia was celebrated by the slaughter of 11,000 wild animals.' Animals fought to the death in Roman circuses were brought from as far as India and Thailand.

Nriagu argues in his *Lead and Lead Poisoning in Antiquity* that the Romans suffered from universal and severe lead pollution. Symptoms of imperial madness, for example, suggest that a large number of Roman emperors were driven insane by the metal. Lead was used for water piping, as tableware and, most dangerously, for the amphorae used to store a favourite and commonly used acidic fish sauce. Goldsmith (1975) has argued more speculatively that a number of linked environmental pollutants and problems led to the collapse of Roman society. Again he speculates on the problems of soil erosion, noting that 'The deserts of North Africa, once the granary of Rome, bear even more eloquent testimony to the destructive agricultural practices of the times.' Their story may be read as a grand example of the ancient sin of *hubris* or overweaning arrogance towards nature, followed by swift and ultimately self-created nemesis. Such an interpretation should not close our minds to the complex relations between society and nature at such an early date. The collapse of Rome may have left soil conservation terraces to decay, allowed 'barbarians' to despoil forests and created other potential dangers. Such speculation aside, it is difficult to imagine a society as hostile to the natural world outside our own age.

J. Donald Hughes

Ecology in ancient Mesopotamia

The earliest cities seem to have shared some of the problems which have become as annoying in their modern counterparts. Babylon, in its day the largest city of the area, had a city wall ten miles long, and even including its suburbs was consequently only of moderate size by modern standards. The evidence of narrow streets and small rooms in houses huddled within the compass of defensible walls tells us that crowding in ancient cities was extreme. Garbage accumulated in the houses, where the dirt floors were continually being raised by the debris, and human wastes were rarely carried further than the nearest street. The water supply, from wells, rivers, and canals, was likely to be polluted. Life expectancy was short, due in part to the high infant mortality. Flies, rodents, and cockroaches were constant pests. Even air pollution was not absent. In addition to dust and offensive odors, the atmosphere filled with smoke on calm days. Even today, in large preindustrial cities such as Calcutta the smoke of thousands of individual cooking fires, in addition to other human activities, produces a definite pall of smoke and dust which seldom dissipates for long. Under these unhealthy conditions, the death rate must have been high in Mesopotamian cities.

As an alluvial land, the Mesopotamian plain had no stone or deposits of metallic ore, and these had to be brought in from the mountains or imported from other countries. While this encouraged the early development of trade, it also meant that the inhabitants had to use the native materials, swamp reeds and clay, for ordinary construction. They built mighty works of baked and unbaked clay bricks – temples, shrines raised on lofty ziggurats, palaces, and walled cities – but the system of canals and other irrigation works is their most remarkable achievement. Incidentally, it is interesting to note that petroleum, the most important natural resource of modern Iraq, did not escape their notice. They used the oil which oozed forth in some places as fuel for their lamps and bitumen for water-proofing their boats.

The attitude of the peoples of Mesopotamia toward nature, from early Sumerian writings down through the Akkadian and Assyrian literatures, is marked by a strong feeling of battle. Nature herself was represented in Mesopotamian mythology as monstrous chaos, and it was only by the constant labors of people and their patron gods that chaos could be overcome and order established. Mesopotamian gods, though they retained their earlier character as nature deities to some extent, were primarily figures which sanctioned order, guarded the cities, upheld government and society, and en-couraged the construction of works which would reproduce on earth the regularity of heaven. The order of heaven was quite apparent to the Mesopotamians, who developed both astrology and astronomy to a high degree and noticed that the motions of the moon and sun, stars and planets are constant and predictable. The labors of the Mesopotamian hero-god Enlil, or Marduk, in slaying the primeval monster of chaos, Tiamat, and creating the world out of her sundered body, reflected the labors of the Mesopotamians them-selves, who built islands in the swamps, raised their cities above the flood plains, and irrigated desert stretches with an orderly series of well maintained canals. They planned their cities so that the major streets of Babylon, for example, crossed each other at right angles in a regular grid pattern, and they laid out their canals in the same way wherever possible. Left to itself, Mesopotamia would have remained a land at the mercy of the capricious river and the merciless sun, in a precarious, shifting balance between tangled marsh and parched desert. But careful works of irrigation conquered sections of that land and won rich sustenance from its basic fertility. Thus a Mesopotamian king could list the construction of a new canal, along with the defeat of his enemies in battle, as the major events of a year of his reign.

Mesopotamians had a well developed sense of the distinction between the tame and the wild, between civilization and wilderness. The proper effort of mankind toward wild things, they believed, is to domesticate them. They did this with such native animals as the donkey and the water buffalo, in addition to keeping the cows, pigs, sheep, and goats already known to their ancestors. They learned the uses of the palm tree and planted it widely. Animals which could not be truly domesticated were hunted – some, like the lion, to extinction. In the *Epic of Gilgamesh*, Enkidu was presented as a man of the wild, a friend and protector of beasts. But when he had been tamed by womanly wiles, his former animal friends feared and fled from him. One of the great feats of Gilgamesh and his now tamed companion was the slaying of Humbaba, the wild protector of the cedar forests in the west, and the seal set upon the defeat of Humbaba was the subjugation of the wilderness; the trees were felled for human use. This ancient legend described an actual ecological event; the cedars of Lebanon, after centuries of exploitation and export to all the surrounding lands, were completely destroyed except for a few small groves, leaving their mountain slopes open to severe erosion. . . .

The cities of Mesopotamia have been desolate mounds for a score of centuries, and only a poor remnant of the 'Fertile Crescent', that green, cultivated area which once arched across the Middle East from Sumeria to Palestine, is visible in photographs taken from space today. This disaster is due not simply to changing climate or the devastating influence of war, though both of these have had important effects. It is a true ecological disaster, due partly to the difficulty of maintaining the canals and keeping them free of silt, but more importantly to the accumulation of salt in the soil. Irrigation water, carried over large areas, was allowed to evaporate with insufficient drainage, and over the centuries in this land of low humidity and scanty rain, the salts carried in by the water concentrated. Such areas had to be abandoned, while new sections were brought under irrigation and cultivation until they in turn suffered the same salinization. A similar process occurs in many places where deserts have been irrigated, as in the Imperial Valley of California today, where the best efforts of modern technology have barely been able to combat it. Salt would not have accumulated in well drained soil, but in Mesopotamia the problem of drainage was especially difficult. Silt and mud carried by the rivers and canals settled out rapidly, so that constant dredging was necessary to keep the canals flowing. Excavated mud piled up along the sides of the canals to a height of thirty feet or more, serving as a barrier to drainage. Eventually the river level was raised well above the surrounding country. The natural remedy to this, flooding and a major shift in the course of the rivers, was catastrophic whenever it happened.

It is significant that the first urban societies were also the first societies to abandon a religious attitude of oneness with nature and to adopt one of separation. The dominant myth and reality in Mesopotamia was the conquest of chaotic nature by divine–human order. Such societies, it must be noted, were ultimately unsuccessful in maintaining a balance with their natural environment.

Ecology in Ancient Civilizations, Albuquerque, N.M.: University of New Mexico Press (1975), 30–3, 34–5.

Plato

Eroded Attica

Now the country was inhabited in those days by various classes of citizens; – there were artisans, and there were husbandmen, and there was also a warrior class originally set apart by divine men. The latter dwelt by themselves, and had all things suitable for nurture and education; neither had any of them anything of their own, but they regarded all that they had as common property; nor did they claim to receive of the other citizens anything more than their necessary food. And they practised all the pursuits which we yesterday described as those of our imaginary guardians. Concerning the country the Egyptian priests said what is not only probable but manifestly true, that the boundaries were in those days fixed by the Isthmus, and that in the direction of the continent they extended as far as the heights of Cithaeron and Parnes; the boundary line came down in the direction of the sea, having the district of Oropus on the right, and with the river Asopus as the limit on the left. The land was the best in the world, and was therefore able in those days to support a vast army, raised from the surrounding people. Even the remnant of Attica which now exists may compare with any region in the world for the variety and excellence of its fruits and the suitableness of its pastures to every sort of animal, which proves what I am saying; but in those days the country was fair as now and yielded far more abundant produce. How shall I establish my words? and what part of it can be truly called a remnant of the land that then was? The whole country is only a long promontory extending far into the sea away from the rest of the continent, while the surrounding basin of the sea is everywhere deep in the neighbourhood of the shore. Many great deluges have taken place during the nine thousand years, for that is the number of years which have elapsed since the time of which I am speaking; and during all this time and through so many changes, there has never been any considerable accumulation of the soil coming down from the mountains, as in other places, but the earth has fallen away all round and sunk out of sight. The consequence is that, in comparison of what then was, there are remaining only the bones of the wasted body, as they may be called, as in the case of small islands, all the richer and softer parts of the soil having fallen away, and the mere skeleton of the land being left. But in the primitive state of the country, its mountains were high hills covered with soil, and the plains, as they are termed by us, of Phelleus were full of rich earth, and there was abundance of wood in the mountains. Of this last the traces still remain, for although some of the mountains now only afford sustenance to bees, not so very long ago there were still to be seen roofs of timber cut from trees growing there, which were of a size sufficient to cover the largest houses; and there were many other high trees, cultivated by man and bearing abundance of food for cattle. Moreover, the land reaped the benefit of the annual rainfall, not as now losing the water which flows off the bare earth into the sea, but, having an abundant supply in all places, and receiving it into herself and treasuring it up in the close clay soil, it let off into the hollows the streams which it absorbed from the heights, providing everywhere abundant fountains and rivers, of which there may still be observed sacred memorials in places where fountains once existed; and this proves the truth of what I am saying.

Such was the natural state of the country, which was cultivated, as we may well believe, by true husbandmen, who made husbandry their business, and were lovers of honour, and of a noble nature, and had a soil the best in the world, and abundance of water, and in the heaven above an excellently attempered climate.

B. Jowett, (ed.), *The Dialogues of Plato*, London: Oxford University Press (1892), 3: 531–3.

Lucretius

The exhausted earth

Even now indeed the power of life is broken, and the earth exhausted scarce produces tiny creatures, she who once produced all kinds and gave birth to the huge bodies of wild beasts. For it is not true, as I think, that the generations of mankind were let down from high heaven by some golden chain upon the fields, nor were they sprung from sea or waves beating upon the rocks, but the same earth generated them which feeds them now from herself. Besides, she of her own accord first made for mortals the bright corn and the luxuriant vineyards, of herself she gave forth sweet fruits and luxuriant pasturage, which now scarce grows great when increased by our toil; and we work out our oxen and the strength of our farmers, we wear out the ploughshare, and then are scarce fed by our fields: so do they grudge their fruits, and ask for work to let them increase. Now the ancient ploughman shaking his head sighs many a time, that the labours of his hands have all come to nothing, and comparing times present with times past often praises the fortunes of his father, and grumbles how the old world full of piety supported life with great ease on a narrow domain; since the man's portion of land was formerly much smaller than it is now. Sadly also the cultivator of the degenerate and shrivelled vine rails at the progress of time and wearies heaven; not comprehending that all things gradually decay, and go to the tomb outworn by the ancient lapse of years.

De Rerum Natura, ed. W. H. D. Rouse, Loeb Classical Library, London: Heinemann (1937), 167–9.

J. Donald Hughes

Ecology in imperial Rome

Erosion caused whole provinces – notably Syria and North Africa – to decline in ability to support their populations. The deposits of rock, sand, and mud washed down by

Mediterranean streams after deforestation are up to thirty feet thick in many places. These materials were washed down from above, where the hillsides, deprived of their natural cover of vegetation, now lost their soil cover as well. Without the forests to hold back water and deliver it more slowly to the areas below, many springs and smaller streams dried up, as some writers at the time reported. During the rains, the water in some of the aqueducts was muddied. Marshlands created by deposits from erosion presented another danger in the form of malaria, which became a widespread disease in Rome early in the second century B.C. after its introduction, possibly from Greece. The lower country near the city of Rome supplies a notorious example of flooding, repeated development of swamps, and the endemic problem of malaria, due to upland erosion. The Romans periodically embarked on ambitious programs for the drainage of marshlands, destroying wildlife habitat as they did so, but many of the new alluvial deposits were too low-lying for efficient drainage, and were never completely reclaimed throughout the period of the Roman Empire. The drainage of the ill famed Pontine Marshes was not permanently accomplished at any time before the twentieth century, because the basic problem of erosion and drainage was chronic.

The use of wild animals by the Romans constituted gross exploitation and may serve as evidence for their general treatment of nature. While their ancestors had lived by hunting, and some poorer people in outlying districts still did so, for the Romans in general hunting was a matter for sport or business. Roman mosaics illustrate the activities of the hunt, showing hunting dogs and falcons. Falconry with hawks and eagles was introduced from Persia. Hunting was a private sport in Rome, although certain privileges, such as hunting lions, were in time reserved for the emperor.

Romans with large estates usually had game parks set aside for themselves and their guests to hunt in, and constructed aviaries for game birds as well. They killed and ate many more species of birds than is common today. Not only the upper classes hunted. Roman peasants hunted smaller animals for food occasionally, and killed the predators who fed on their herds. The wolf, maternal symbol of the founding of Rome, was extirpated from the more thickly settled areas.

Rome demanded wild animals and their products from the provinces of the empire and beyond. Ivory from Africa and India was used in works of art, including huge statues of ivory and gold, and was inlaid in furniture of every kind. Ivory writing tablets, desks, spoons, and other objects were popular among the Romans. Wild animal skins were used for clothing and furnishings, and feathers of ostriches and other birds served for decorations on military uniforms and elsewhere.

But the entertainment industry of Rome, which included the display and killing of countless animals, was more wasteful and destructive of wildlife than any other feature of Roman culture. The shows put on in amphitheaters for the amusement of the people included animals who could do tricks and take part in pageants and plays to indulge Roman tastes, which ran heavily to sex and violence. Some of the rarer animals were merely shown as curiosities, but more often they were mutilated and killed. *Venationes*, or hunts, in which armed men, sometimes even the Praetorian cavalry, chased and killed animals, constituted a major part of the shows, and are often the subjects of Roman mosaics and paintings. Fights were staged between goaded and crazed animals – between a bull and a rhinoceros, for example. Unarmed or poorly armed people, usually condemned criminals, including those guilty of the 'crime' of belonging to an illegal religion such as Christianity, were exposed to starving animals such as lions, leopards, and bears. Special cages were constructed under the amphitheater, complete with elevators and ramps to bring the beasts up to the arena without endangering the attendants too much. Romans of every social level, from the emperor to the common people, attended the games and enjoyed a spectacle which can only be called sadistic, and Roman writers, with rare exceptions, describe the games with approval.

The variety of wild animals and the vast numbers killed must be of the greatest concern in this book. Only the larger animals were used in the arena, since they had to be visible to thousands of people at once. But all kinds of large mammals, reptiles, and birds were imported from the empire, Europe, Africa, and Asia as far as India and even Thailand. Elephants were first seen in Rome in 275 B.C., after Pyrrhus of Epirus brought them to Italy in his military campaign. Ninety years later, ostriches, leopards, and lions were seen, and, in the last century of the Republic, hippopotamuses, crocodiles, and rhinoceroses made their entrance from Egypt, and Caesar sent a lynx from Gaul. Augustus displayed tigers from India, and Nero showed polar bears catching seals.

The numbers of animals killed are phenomenal, mounting into the hundreds in a single day. Augustus had 3,500 animals killed in twenty-six *venationes*. At the dedication of the Colosseum under Titus, 9,000 were destroyed in 100 days, and Trajan's conquest over Dacia was celebrated by the slaughter of 11,000 wild animals.

Since there were many amphitheaters throughout the Roman Empire, the demand for wild animals was enormous and was supported by an extensive, organized business for hunting, capture, and transportation. Many found employment in this enterprise, which was far from easy, since the beasts had to be kept in good condition in pits, nets, or cages from the time they were captured until they were delivered, as in the case of the city of Rome, to the menagerie or *vivarium* outside the Praenestine Gate. For the most part, it was a private business on which the Roman government levied an import tax of 2½ per cent, and many of the animals went to private parks. Roman officials were deeply involved in the trade, however, and soldiers were sometimes used to round up the animals. Animals specifically captured for the emperor had their food and other needs requisitioned from the towns through which they passed – not a small charge for some towns on the usual routes, considering the size of the animals and their number. Only the emperor was permitted to own elephants, and had a special officer, the *procurator ad elephantos*, to keep the imperial herd at Ardea, but in Republican times at least one private citizen had made a custom of impressing his friends by riding an elephant when he went out to dinner. Emperors often kept impressive animals of various kinds in their palaces, and rich citizens sometimes had tame lions or other animals as pets. More ordinarily, they kept house cats, those terrors of mice and garden songbirds, which they spread from Egypt through Western Europe. Granted all of this, it is surprising that the Romans did not follow the lead of the Hellenistic Greeks in establishing zoos. Certainly they directed little scientific study toward animals, although Galen and other physicians did take the opportunity of viewing the internal anatomy opened to their view by the mutilations of the arena.

The result of the Roman procurement of animals was the extinction of larger mammals, reptiles, and birds from the areas most accessible to the professional hunters and trappers. The Romans themselves boasted of this, pointing out that they were removing dangers to man and his agriculture. But they also exhausted the hunting grounds of North Africa, where the elephant, rhinoceros, and zebra became extinct. The hippopotamus was extirpated from the lower Nile, and lions from Thessaly, Asia Minor, and parts of Syria. Tigers disappeared from Hyrcania in northern Iran, the closest source to Rome. Of course, these creatures were hunted and killed by other people and for other reasons too, but the Roman demand for animals effectively aided in making large wild animals rare or extinct in the entire Mediterranean Basin.

Ecology in Ancient Civilizations, Albuquerque, N.M.: University of New Mexico Press (1975), 102–6.

J. Nriagu

Lead and lead poisoning in antiquity

Lead and Lead Poisoning in Antiquity by J. Nriagu is reprinted by permission of John Wiley & Sons, Inc.

Lead possess unique properties that were to make it one of the most useful industrial metals in Greek and Roman times. Because of its corrosion resistance and formability, it was used extensively in plumbing, architecture, shipbuilding, and for stationery. Its density and malleability made it attractive for making plummets, sinkers, and standard weights. Its low melting point – further reduced by the addition of tin – ensured its use as solder since very ancient times. The addition of lead to bronzes makes them easier to cast, and it results in what Pliny called 'statue metal'. Because of its atomic configuration, large amounts of lead could be added to silicate glasses, where it functioned as an opacifier or a colorant. Several compounds of lead are brilliantly colored and were valued as pigments since Palaeolithic times. Even the toxic properties of lead were put to good use. The utilization of lead reached such an impressive level during the period of the Roman Empire that lead is often referred to as a 'Roman metal'. The use of lead in the Roman Empire exceeded 550 g per person per year. The cumulative worldwide production of lead from the earliest times to the fall of the Roman Empire has been estimated to be about 40 million tons. . . .

Because of the toxic properties of lead, the price of its utilization has often been high. I believe that the first person to commercialize metallic lead was probably poisoned by the lead fumes from his or her kiln or furnace, and undoubtedly generations of artisans throughout antiquity who worked with this dangerous metal received the same rude treatment. It is estimated that the number of workers who were occupationally exposed to lead during the period of the Roman Empire was over 140,000 per year. Considerably higher fractions of lead-using populations were exposed to lead contamination in their food and drink. The Romans, for example, preserved their fruits and vegetables with lead salts, cooked their foods in leaden pots, and commonly assuaged their 'sweet tooth' with the sugar of lead (*saccharum saturni*, or lead acetate). They added lead to their wines to stop further fermentation, to impart color or bouquet, or to blunt the acidity of an erratic brew. Their water was delivered in lead pipes, while saturnine cosmetics and medicaments were common and quite popular. With such overexposure to lead, we find frequent literary references to epidemics of plumbism and saturnine gout among the members of the

Roman aristocracy. Other historiographic evidence indicating that lead poisoning caused reproductive failure of the ruling oligarchy includes the high incidence of sterility, the alarming rates of stillbirths, and the well known mental incompetence of the progeny of the aristocrats. Indeed, the psychological profiles of the emperors and usurpers who reigned between 50 B.C. and A.D. 250 suggest that the majority of them probably suffered from lead poisoning. A full documentation of the hypothesis that the Roman Empire decline as a result of plumbism among members of the aristocratic oligarchy is provided in the final chapter.

... roughly two thirds (eighteen out of thirty) of the emperors and usurpers who reigned between 30 B.C. (Augustus) and A.D. 220 (Elegabalus) had a predilection for lead-tainted Apician *entrées* and Columellan wine blends. This figure basically is confirmed by Celsus (4.24), who observed that most of the Roman emperors suffered from gout. It may be assumed that a large number of their well placed cohorts acquired similar tastes. This continuing parade of unproductive ('aped' by lead) rulers and administrators throughout this period of the Roman Empire no doubt contributed immeasurably to social disequilibrium and economic disaster.

Pundits obviously need to be reminded that hundreds of reasons have been suggested for the collapse of the Roman Empire. In the classic history *The Decline and Fall of the Roman Empire*, Edward Gibbon (1776–88) listed over two dozen causes of the decline. Reasons often mentioned include the following: economic stagnation, demographic decline including manpower shortages, over-exploitation of the masses and natural resources, the inflexibility, or the monolithic nature, of the socio-political system, systemic senescence, and so on (see Bark 1958; Remondon 1964; Jones 1966; Lot 1966; Perowne 1966; Macmullen 1967; Vogt 1967; Jordon 1971; Grant 1976; Kagan 1978). In the final analysis, however, Rome fell to easy-to-identify invasions from outside and weaknesses from within. The present study of the production and use of lead – often referred to as the 'Roman' metal – by Roman culture strongly implicates lead poisoning of the aristocratic oligarchy as one of the principal, probable causes of the internal weakness. Kobert (1909) and Gilfillan (1965) had previously reached the same basic conclusion.

Lead and Lead Poisoning in Antiquity, London: Wiley (1983), vii–viii, 414–15.

CHAPTER 3

The origins of environmental danger

J. Evelyn, Fumifugium

A. H. Hassall, Food additives

F. Engels, The great cities

M. Somerville, Necessities and enjoyments . . .

J. Muir, The destruction of the redwoods

Introduction

> I durst not laugh for fear of opening my lips and receiving the bad air.
> (Shakespeare, *Julius Caesar*, Act 1, (Scene 1, vv. 249–50))

Seemingly contemporary environmental problems have apparently existed in the past and fuelled ecological concern. The Romans, as we have seen, undoubtedly suffered from lead poisoning. Air pollution may have been an unpleasant reality for the inhabitants of ancient Mesopotamia. Poor planning decisions and industrial toxic waste blighted medieval communities. Dilemmas that seemed unimaginable before the Second World War have been speculated upon far earlier. For example, a Swedish scientist, Svante Arrhenius (1859–1927) noted that coal burning was 'evaporating our coal mines into the air'. Doubling the amount of natural carbon dioxide in the atmosphere would cause a 5°C rise in temperature worldwide. The ice caps might melt and sea levels would rise; the greenhouse effect had been described. A century earlier the Egyptologist Jean-Baptiste Joseph Fourier (1768–1830), who developed the theory of heat conduction, also speculated on such a 'hothouse' effect. The physicist Frederick Soddy (1877–1956) worried about the ill effects inherent in exploiting the atom in the 1920s, decades before the discovery of nuclear power or the invention of the atom bomb (Martinez-Alier 1990: xv).

The first law against air pollution was enacted in Britain in reign of Richard II during the thirteenth century. It did little good, and by the 1560s London was bathed in sulphurous smoke. Here we find John Evelyn (1620–1706) describing the noxious effects of this pollution, tracing its cause to industry and suggesting laws to banish polluters from the city. Evelyn campaigned against effluent, vigorously supported tree planting and even propounded a vegetarian diet, yet, like the conservationists of Theodore Roosevelt's era such as Pinchot, his approach to nature was essentially managerial. He sought to find ways of exploiting nature more efficiently rather than rejecting such assaults, lacked a radical Green reverence for nature and as a supporter of the monarchy was no rebel. His essay *Fumifugium* is, though, an early, if not the earliest, example of environmental lobbying. He describes the problem, calculates the cost in terms of poor health, damage to buildings and the destruction of gardens; presents a solution and draws up a programme of legislation for Parliament.

Adulterations Detected was the result of a long (and partially successful) campaign to control food additives, including artificial colouring, in Victorian England. The debate seems to have changed little a century later, with manufacturers continuing to argue that additives mean cheaper food and are popular with consumers. Not all forms of adulteration took the form of exotic and deadly chemicals – sawdust and chalk were regularly added to bread – however, the use of toxic heavy metals as food colouring agents is horrifyingly catalogued here. *Adulterations Detected* was designed to act as a scientific manual illustrating how the expert might discern and act against illegal toxic additives.

Friedrich Engels (1820–95) study *The Condition of the Working Class in England* is of twofold interest. It indicates that, despite the efforts of Evelyn and others, pollution problems had multiplied with the coming of the industrial revolution. Equally it indicates how environmental concern helped propel Engels towards the communist movement and his lifelong collaboration with Karl Marx (Carver 1983: 3–4). Written when he was a very young man, the volume links social critique with environmental awareness. Next we find the geographer Mary Somerville (1780–1872) commenting on the likely extinction of animals that modern wildlife campaigners have fought long and hard to preserve. Her pessimism may still, despite such efforts, be realized. Finally in 1901 the deep ecologist John Muir recounts how the American redwoods were swept away in a process hauntingly similar to the modern destruction of the rain forests. While the greenhouse effect, nuclear power and the supposedly thinning ozone layer, together with the dubious fruits of biotechnology, provide worries for our present society, many other ecological threats have a longer pedigree. Perhaps modern industrial development has merely amplified long-standing trends.

John Evelyn

Fumifugium

… her *Inhabitants* breathe nothing but an impure and think Mist, accompanied with a fuliginous and filthy vapour, which renders them obnoxious to a thousand inconveniences, corrupting the *Lungs*, and disordering the entire habit of their Bodies; so that *Catharrs*, *Phthisicks*, *Coughs* and *Consumptions*, rage more in this one City, than in the whole Earth besides.

I shall not here much descant upon the Nature of *Smoakes*, and other Exhalations from things burnt, which have obtained their several *Epithetes*, according to the quality of the Matter consumed, because they are generally accounted noxious and unwholesome; and I would not have it thought, that I doe here *Fumos vendere*, as the word is, or blot paper with insignificant remarks: It was yet haply no inept derivation of that *Critick*, who took our *English*, or rather, *Saxon* appellative, from the *Greek* word ομύχω *corrumpo* and *exuro*, as most agreeable to its destructive effects, especially of what we doe here so much disclaim against, since this is certain that of all the common and familiar materials which emit it, the immoderate use of, and indulgence to *Sea-coale* alone in the City of *London*, exposes it to one of the fowlest Inconveniencies and reproches, than can possibly befall so noble, and otherwise incomparable City: And that, not from the *Culinary* fires, which for being weak, and lesse often fed below, if with such ease dispelled and scattered above, as it is hardly at all discernible, but from some few particular Tunnells and Issues, belonging only to *Brewers, Diers, Lime-burners, Salt,* and *Sope-boylers,* and some other private Trades, *One* of whose *Spiracles* alone, does manifestly infect the *Aer,* more than all the Chimnies of *London* put together besides. And that this is not the least *Hyperbolie,* let the best of Judges decide it, which I take to be our senses: Whilst these are belching it forth their sooty jaws, the City of *London* resembles the face rather of *Mount Ætna,* the *Court of Vulcan, Stomboli,* or the Suburbs of *Hell,* than an Assembly of Rational Creatures, and the Imperial seat of our incomparable *Monarch.* For when in all other places the *Aer* is most Serene and Pure, it is here Ecclipsed with such a Cloud of Sulphure, as the Sun itself, which gives day to all the World besides, is hardly able to penetrate and impart it here; and the weary *Traveller,* at many Miles distance, sooner smells, than sees the City to which he repairs. This is that pernicious Smoake which sullyes all her Glory, superinducing a sooty Crust or Furr upon all that it lights, spoyling the moveables, tarnishing the Plate, Gildings and Furniture, and corroding the very Iron-bars and hardest Stones with those piercing and acrimonious Spirits which accompany its Sulphure; and executing more in one year, than exposed to the pure *Aer* of the Country it could effect in some hundreds. . . .

It is this horrid Smoake which obscures our Churches, and makes our Palaces look old, which fouls our Clothes, and corrupts the Waters, so as the very Rain, and refreshing Dews which fall in the several Seasons, precipitate this impure vapour, which, with its black and tenacious quality, spots and contaminates whatever is exposed to it. . . .

It is this which scatters and strews about those black and smutty *Atomes* upon all things where it comes, insinuating itself into our very secret *Cabinets,* and most precious *Repositories*: Finally, it is this which diffuses and spreads a Yellownesse upon our choycest Pictures and Hangings: which does this mischief at home, is *Avernus* to *Fowl,* and kills out *Bees* and *Flowers* abroad, suffering nothing in our Gardens to bud, display themselves, or ripen: so as our *Anemonies* and many other choycest Flowers, will by no Industry be made to blow in *London,* or the Precincts of it, unlesse they be raised on a *Hot-bed,* and governed with extraordinary Artifice to accellerate their springing; imparting a bitter and ungrateful Tast to those few wretched *Fruits,* which never arriving to their desired maturity, seem, like the *Apples* of *Sodome,* to fall even to dust, when they are but touched. . . .

We know (as the *Proverb* commonly speaks) that, as *there is no Smoake without Fire*; so neither is there hardly any *Fire* without *Smoake,* and that the materials which burn clear are very few, and but comparatively so tearmed: That to talk of serving this vast City (though *Paris* as great, be so supplied) with *Wood,* were madnesse; and yet doubtlesse it were possible, that much larger proportions of Wood might be brought to *London,* and sold at easier rates, if that were diligently observed, which both our *Laws* enjoyn, as faisible and practised in other places more remote, by Planting and preserving of *Woods* and *Copses,* and by what might by Sea, be brought out of the *Northern Countries,* where it so greatly abounds, and seems inexhaustible. But the *Remedy* which I would propose, has nothing in it of this difficulty, requiring only the Removal of such *Trades,* as are manifest *Nuisances* to the City, which, I would have placed at farther distances; especially, such as in their Works and Fournaces use great quantities of *Sea-Coale,* the sole and only cause of those prodigious Clouds of *Smoake,* which so universally and so fatally infest the *Aer,* and would in no City of *Europe* be permitted, where Men had either respect to Health or Ornament. Such we named to be *Brewers, Diers, Sope* and *Salt-boylers, Lime-burners,* and the like. . . .

I propose therefore, that by an *Act* of this present *Parliament*, this infernal *Nuisance* be reformed; enjoyning, that all those *Works* be removed five or six miles distant from *London* below the River of *Thames*; I say, five or six Miles, or at the least so far as to stand behind that *Promontary* jetting out, and securing *Greenwich* from the pestilent *Aer* or *Plumstead-* Marshes: because, being placed at any lesser Interval beneath the *City*, it would not only prodigiously infect that his *Majesties* Royal Seat (and as *Barclay* calls it) *pervetusta Regum Britannicorum domus*; but during our nine Months *Etesians* (for so we may justly name our tedious Western-winds) utterly darken and confound one of the most princely, and magnificent Prospects that the World has to shew: Whereas, being seated behind that Mountain, which seems to have been thus industriously elevated, no winds, or other accident whatever can force it through that solid obstacle; and I am perswaded, that the heat of these Works, mixing with the too cold and uliginous vapours which perpetually ascend from these Fenny Grounds, might be a means of rendring that *Aer* far more healthy then now it is; because it seems to stand in need of some powerful drier; but which *London*, by reason of its excellent scituation, does not at all require.

Fumifugium (1661), Oxford: Ashmolean Reprints, (1930), 18–21, 34, 36.

A. H. Hassall, M.D.

Food additives

One reason assigned in defence of any adulterations is that they are practised in obedience to the wishes and tastes of the public ... [thus] ... the practice of colouring the red sauces, potted meats, and fish with bole Armenian; cheese with amatto; pickles, bottled fruits, and vegetables, with copper; and sugar confectionary with various pigments consisting of salts of arsenic, copper, zinc, and antimony, is excused....

When, therefore, the manufacturer or seller defends any particular admixture, or adulteration, on any of the pleas referred to, namely, that it is practised to suit the public taste, that it is an improvement, or that it is necessary in order to make the article keep, we would advise our readers to look well into the matter for themselves; they will be almost sure to find something wrong, some fallacy at the bottom of these statements, – they will too often find that this pretended regard for the wishes and tastes of the public resolves itself into a question of gain to the manufacturer or trader....

It may so happen, and it doubtless does sometimes occur, that the same person, in the course of a single day, receives into his stomach some eight or ten of the articles above enumerated. Thus, with the potted meats and fish, anchovies, red sauces or cayenne, taken at breakfast, he would consume more or less bole Armenian, Venetian red, red lead, or even bisulphate of mercury. At dinner, with his curry or cayenne, he would run the chances of a second dose of lead or mercury; with the pickles, bottled fruits and vegetables, he would be nearly sure to have copper administered to him; while if he partook of bon bons at dessert, there is no telling what number of poisonous pigments he might consume. Again, in his tea, of mixed or green, he would certainly not escape without the administration of a little Prussian blue, and it might be worse things: if he were a snuff-taker, he would be pretty sure to be putting up his nostrils, from time to time, small quantities of either some ferruginous earth, bichromate of potash, cromate of lead, or red lead: finally, if he indulged himself with a glass or so of grog before going to bed, he would incur the risk of having the coats of his stomach burned and irritated with tincture of capsicum or essence of cayenne. If an invalid, his condition would be still worse; for then, in all probability, he would be deprived of much of the benefit of the skill of his physician through the dilution and sophistication to which the remedies administered for his relief were subjected.

Adulterations Detected, or, Plain Instructions for the Discovery of Frauds in Food and Medicine. London: Longman (1861), 4, 5, 22.

Friedrich Engels

The great cities

The centralisation of population in great cities exercises of itself an unfavourable influence; the atmosphere of London can never be so pure, so rich in oxygen, as the air of the country; two and a half million pairs of lungs, two hundred and fifty thousand fires, crowded upon an area three to four miles square, consume an enormous amount of oxygen, which is replaced with difficulty, because the method of building cities in itself impedes ventilation. The carbonic acid gas, engendered by respiration and fire, remains in the streets by reason of its specific gravity, and the chief air current passes over the roofs of the city. The lungs of the inhabitants fail to receive the due supply of oxygen, and the consequence is mental and physical lassitude and low vitality. For this reason, the dwellers in cities are far less exposed to acute, and especially to inflammatory, affections than rural populations, who live in a free, normal atmosphere; but they suffer the more from chronic affections. And if life in large cities is, in itself, injurious to health, how great must be the harmful influence of an abnormal atmosphere in the working-people's quarters, where, as we have seen, everything combines to poison the air. In the country, it may, perhaps, be comparatively innoxious to keep a dung-heap adjoining one's dwelling, because the air has free ingress from all sides; but in the midst of a large town, among closely built lanes and courts that shut out all movement of the atmosphere, the case is different. All putrefying vegetable and animal substances give off gases decidedly injurious to health, and if these gases have no free way of escape, they inevitably poison the atmosphere. The filth and stagnant pools of the working-people's quarters in the great cities have, therefore, the worst effect upon the public health, because they produce precisely those gases which engender disease; so, too, the exhalations from contaminated streams. But this is by no means all. The manner in which the great multitude of the poor is treated by society to-day is revolting. They are drawn into the large cities where they breathe a poorer atmosphere than in the country; they are relegated to districts which, by reason of the method of construction, are worse ventilated than any others; they are deprived of all means of cleanliness, of water itself, since pipes are laid only when paid for, and the rivers so polluted that they are useless for such purposes; they are obliged to throw all offal and garbage, all dirty water, often all disgusting drainage and excrement into the streets, being without other means of disposing of them; they are thus compelled to infect the region of their own dwellings. Nor is this enough. All conceivable evils are heaped upon the heads of the poor. If the population of great cities is too dense in general, it is they in particular who are packed into the least space. As though the vitiated atmosphere of the streets were not enough, they are penned in dozens into single rooms, so that the air which they breathe at night is enough in itself to stifle them. They are given damp dwellings, cellar dens that are not waterproof from below, or garrets that leak from above. Their houses are so built that the clammy air cannot escape. They are supplied bad, tattered, or rotten clothing, adulterated and indigestible food.

The Conditions of the Working Class (1844), London: Allen & Unwin (1920), 96–8.

Mary Somerville

Necessities and enjoyments . . .

Man's necessities and enjoyments have been the cause of great changes in the animal creation, and his destructive propensity of still greater. A farmer sees the rooks pecking a little of his grain, or digging at the roots of the springing corn, and poisons all in his neighbourhood. A few years after he is surprised to find his crop destroyed by grubs. The works of the Creation are nicely balanced, and man cannot infringe the laws of equilibrium with impunity. Insects would become torments were they not kept in check by birds. Animals soon acquire a dread of man, which becomes instinctive and hereditary: in newly discovered uninhabited countries birds and beasts are so tame as to allow themselves to be caught; whales scarcely got out of the way of ships that first navigated the Arctic Ocean, but now they universally have a dread of the common enemy: whales and seals have been extirpated in various places: sea fowl and birds of passage are not likely to be extinguished, but many land animals and birds are disappearing before the advance of civilisation. Drainage, cultivation, cutting down of forests, and even the introduction of new plants and animals, destroy some of the old and alter the relations between those that remain. The inaccessible cliffs of the Himalayas and Andes will afford a refuge to the eagle and condor, but the time will come when the mighty forests of Bhutan, of the Amazon and Orinoco, will disappear with the myriads of their joyous inhabitants.

This time is at present, however extremely remote. It is true that the process of occupying and replensing the earth, by the most active and restless of the species, is advancing with accelerating rapidity. All the remaining central part of one of the largest continents of the earth is now fast filling up with civilised inhabitants. It is mapped into states, territories, and counties, and a line of iron road, over which pants the irresistible locomotive, spans the wide region from ocean to ocean. A vast migration streams from the shores of Europe to this rapidly advancing new continent. In the other hemisphere, too, Australia and New Zealand are fast becoming peopled by the same process. But it will be long indeed before the whole of the fertile land of the earth is occupied by a progressive, civilised people. The great interior plains of South America alone would sustain a larger population than China, and at present their inhabitants are only in the proportion of one to four square miles of territory. The enervating climate of the country will render the settlement of this great region by the Caucasian race an extremely slow process. The same may be said of the continent of Africa, which possesses a wide region of comparatively healthy upland in its broad interior, but which offers, in the deadly climate of its coast lands, a formidable obstacle to the occupation of the better tracts by a higher race than has now the possession of the land. The natural forest clothing of as great portion of these continents, with its myriads of curious and beautiful forms of animal and vegetable life, is destined probably to disappear, and man alone, with his cultivated plants and domesticated animals, will occupy the place of the natural tenants: but there will remain many a rugged valley in the lofty mountain ranges, and many a league of inaccessible swamp, to serve as a refuge for a large portion of the native faunas and floras. We may hope, moreover that, with the increase of wealth, knowledge, and refinement, which happily seems a secure prospect for the long vistas of the future, man will endeavour to preserve the equilibrium which exists in the meteorological forces and vital conditions of countries, when in their natural state, by fostering a due proportion of woodland, and thus save from extinction the myriad beauteous forms of life which have shared with him the inhabitance of this wonderful earth.

Physical Geography (1848), London: John Murray (1877), 504–5.

J. Muir

The destruction of the redwoods

Under the timber and stone act of 1878, which might as well have been called the 'dust and ashes act', any citizen of the United States could take up one hundred and sixty acres of timber land, and by paying two dollars and a half an acre for it obtain title. There was some virtuous effort made with a view to limit the operations of the act by requiring that the purchaser should make affidavit that he was entering the land exclusively for his own use, and by not allowing any association to enter more than one hundred and sixty acres. Nevertheless, under this act wealthy corporations have fraudulently obtained title to from ten thousand to twenty thousand acres or more. The plan was usually as follows. A mill company, desirous of getting title to a large body of redwood or sugar-pine land, first blurred the eyes and ears of the land agents, and then hired men to enter the land they wanted, and immediately deed it to the company after a nominal compliance with the law; false swearing in the wilderness against the government being held of no account. In one case which came under the observation of Mr. Bowers, it was the practice of a lumber company to hire the entire crew of every vessel which might happen to touch at any port in the redwood belt, to enter one hundred and sixty acres each and immediately deed the land to the company, in consideration of the company's paying all expenses and giving the jolly sailors fifty dollars apiece for their trouble.

By such methods have our magnificent redwoods and much of the sugar-pine forests of the Sierra Nevada been absorbed by foreign and resident capitalists. Uncle Sam is not often called a fool in business matters, yet he has sold millions of acres of timber land at two dollars and a half an acre on which a single tree was worth more than a hundred dollars. But this priceless land has been patented, and nothing can be done now about the crazy bargain. According to the everlasting law of righteousness, even the fraudulent buyers at less than one per cent of its value are making little or nothing, on account of fierce competition. The trees are felled, and about half of each giant is left on the ground to be converted into choice lumber and sold to citizens of the United States or to foreigners: thus robbing the country of its glory and impoverishing it without right benefit to anybody, – a bad, black business from beginning to end.

This redwood is one of the few conifers that sprout from the stump and roots, and it declares itself willing to begin immediately to repair the damage of the lumberman and also that of the forest-burner. As soon as a redwood is cut down or burned it sends up a crowd of eager, hopeful shoots, which, if allowed to grow, would in a few decades attain a height of a hundred feet, and the strongest of them would finally become giants as great as the original tree. Gigantic second and third growth trees are found in the redwoods, forming magnificent temple-like circles around charred ruins more than a thousand years old. But not one denuded acre in a hundred is allowed to raise a new forest growth. On the contrary, all the brains, religion, and superstition of the neighbourhood are brought into play to prevent a new growth. The sprouts from the roots and stumps are cut off again and again, with zealous concern as to the best time and method of making death sure. In the clearings of one of the largest mills on the coast we found thirty men at work, last summer, cutting off redwood shoots 'in the dark of the moon', claiming that all the stumps and roots cleared at this auspicious time would send up no more shoots. Anyhow, these vigorous, almost immortal trees are killed at last, and black stumps are not their only monuments over most of the chopped and burned areas.

The redwood is the glory of the Coast Range. It extends along the western slope, in a nearly continuous belt about ten miles wide, from beyond the Oregon boundary to the south of Santa Cruz, a distance of nearly four hundred miles, and in massive, substantial grandeur and closeness of growth surpasses all the other timber woods of the world. Trees from ten to fifteen feet in diameter and three hundred feet high are not uncommon, and a few attain a height of three hundred and fifty feet or even four hundred, with a diameter at the base of fifteen to twenty feet or more, while the ground beneath them is a garden of fresh, exuberant ferns, lilies, gaultheria, and rhododendron. This grand tree, Sequoia sempervirens, is surpassed in size only by its near relative, Sequoia gigantea, or Big Tree, of the Sierra Nevada, if, indeed, it is surpassed. The sempervirens is certainly the taller of the two. The gigantea attains a greater girth, and is heavier, more noble in port, and more sublimely beautiful. These two Sequoias are all that are known to exist in the world, though in former geological times the genus was common and had many species. The redwood is restricted to the Coast Range, and the Big Tree to the Sierra.

As timber the redwood is too good to live.

Our National Parks, Boston, Mass.: Houghton Mifflin (1901), 347–50.

CHAPTER 4

Theories of breakdown

G. Agricola, *In defence of mining the earth*

F. Bacon, *New Atlantis*

A. Jackson, *Humanity has often wept*

J. F. Cooper, *Notions of the Americans*

E. Bellamy, *Looking backward*

The Webbs, *Soviet communism: a new civilization*

Introduction

> Christianity may also have contributed to science a technological or mechanistic picture of nature. By denying to nonhuman entities a soul or indwelling spirit, Christianity helped reduce man's perception of nature to the status of a mechanical contrivance.... All is fashioned according to wholly rational, intelligible design that is imposed on chaos. And all will be destroyed when it has ceased to serve the purposes of its Creator. Neither our science nor our technology, it seems likely, would have been possible without the mental preparation afforded by this peculiar myth of origin.
>
> (Worster 1991: 29)

Whilst threads of ecological thought have existed for millennia amongst varied human societies, such ideas have often remained peripheral to the accepted wisdom. In late eighteenth century, at the same time that Romantics like Blake and Goethe wrote their poetry, industrialism was proceeding without any concern for the natural world, and far from peaceful colonial empires built up by France, Britain and other European countries were exploiting people and the environment on a global scale. Carlyle, Marx, Morris and Ruskin, who in their different ways criticized capitalist expansion, were neither the loudest nor the most influential voices to be heard.

In looking for the root causes of what may be crudely termed 'anti-ecological' attitudes, Greens can be subdivided between those who seek economic explanations, e.g. capitalism, population increase, and those who prefer to look at belief systems. Those who prefer ideas to economics can be further subdivided into a number of distinct camps. Capra (1983) argues, for example, that a reductionist understanding of the world, which splits nature up into small components that can be easily manipulated, is the cause of crisis. Other claim that Christianity, with its call in Genesis (1: 26–31) for humanity to 'be fruitful and multiply', so as to 'have dominion over every living thing that moveth upon the earth', is the real culprit (White 1967). Aristotle, who held that nature was merely a resource to be used for human benefit, has also been seen as a source of an anti-ecological ideology (Hughes 1975a). Sohn-Rethel (1978) believes from a Marxist perspective that the growth of industry gave rise to the reductionist canon of Newton and Descartes, rather than vice versa. Adorno and Horkheimer argued that the reason and science developed out of the eighteenth-century Enlightenment were used as tools to dominate both people and nature. The extracts in this chapter illustrate the range of responses to the environment that stress the need for humanity to dominate and subdue it. These include, for example, defences of economic rationality, centralized planning, reductionist scientific method and other broadly hostile notions.

Georgius Agricola (1494–1555) sought to show how scientific principles might be applied to capture Earth's mineral wealth. His manual defends the mining industry from vocal sixteenth-century environmental criticism, the pollution caused by mining is noted, but the 'Gaia hypothesis', which states that the Earth is a living creature which should not be assaulted is rejected. God, argues Agricola, provided for human needs by placing metals in the ground.

Francis Bacon (1561–1626), Lord Chancellor to James I of England, posthumously a target of the poet Blake's considerable spleen, represents the voice of reductionist science harnessed to industrial expansion and state power. *Saloman's House* is a palace of experiments put in place to increase the power of the elite over the majority and the environment. Amongst other 'techniques', vivisection is defended.

Andrew Jackson (1767–1845), a US President, and James Fenimore Cooper (1789–1851), an American novelist famed for *The Last of the Mohicans*, both writing around 1830, stress that wild nature and wild humanity (in the form of the native American) have been cleared away to allow civilization to flourish. His defence of the 'present policy', which 'although regrettable' in his eyes is an inevitable element of sound economic development, sounds like any twentieth-century progressophile's defence of the loss of an aboriginal tribe resettled, or rain forest cleared for new mineral exploration. Cooper describes the fruits of the industrial and military might prayed for by Bacon.

Edward Bellamy (1850–98), a late nineteenth-century socialist, in his utopian novel *Looking Backward*, also defends notions of progress, inevitable human growth and the benefits of a consumer economy. The book inspired William Morris to write his eco-utopian *News from Nowhere* in disgust. Finally, Beatrice and Sidney Webb robustly defend Stalin's plans to remould both the Russian people and their environment. Ironically the Fabian Society, of which the Webbs were leading members, had its origins in the Fellowship of New Life, founded in the 1880s as a transcendentalist body drawing upon the, often Green, beliefs of Thoreau, Tolstoy and Whitman (Gould 1988).

Georgius Agricola

In defence of mining the earth

The critics say further that mining is a perilous occupation to pursue, because the miners are sometimes killed by the pestilential air which they breathe; sometimes their lungs rot away; sometimes the men perish by being crushed in masses of rock; sometimes, falling from the ladders into the shafts, they break their arms, legs, or necks; and it is added there is no compensation which should be thought great enough to equalize the extreme dangers to safety and life. These occurrences, I confess, are of exceeding gravity, and moreover, fraught with terror and peril, so that I should consider that the metals should not be dug up at all, if such things were to happen very frequently to the miners, or if they could not safely guard against such risks by any means. Who would not prefer to live rather than to possess all things, even the metals? For he who thus perishes possesses nothing, but relinquishes all to his heirs. But since things like this rarely happen, and only in so far as workmen are careless, they do not deter miners from carrying on their trade any more than it would deter a carpenter from his, because one of his mates has acted incautiously and lost his life by falling from a high building. I have thus answered each argument which critics are wont to put before me when they assert that mining is an undesirable occupation, because it involves expense with uncertainty of return, because it is changeable, and because it is dangerous to those engaged in it.

Now I come to those critics who say that mining is not useful to the rest of mankind because, forsooth, gems, metals, and other mineral products are worthless in themselves. This admission they try to extort from us, partly by arguments and examples, partly by misrepresentations and abuse of us. First, they make use of this argument: 'The earth does not conceal and remove from our eyes those things which are useful and necessary to mankind, but on the contrary, like a beneficent and kindly mother she yields in large abundance from her bounty and brings into the light of day the herbs, vegetables, grains, and fruits, and the trees. The minerals on the other hand she buries far beneath in the depths of the ground; therefore, they should not be sought. But they are dug out by wicked men who, as the poets say, are the products of the Iron Age.' . . .

But besides this, the strongest argument of the detractors is that the fields are devastated by mining operations, for which reason formerly Italians were warned by law that no one should dig the earth for metals and so injure their very fertile fields, their vineyards, and their olive groves. Also they argue that the woods and groves are cut down, for there is need of an endless amount of wood for timbers, machines, and the smelting of metals. And when the woods and groves are felled, then are exterminated the beasts and birds, very many of which furnish a pleasant and agreeable food for man. Further, when the ores are washed, the water which has been used poisons the brooks and streams, and either destroys the fish or drives them away. Therefore the inhabitants of these regions, on account of the devastation of their fields, woods, groves, brooks and rivers, find great difficulty in procuring the necessaries of life, and by reason of the destruction of the timber they are forced to greater expense in erecting buildings. Thus, it is said, it is clear to all that there is greater detriment from mining than the value of the metals which the mining produces. . . .

. . . And very many vehemently praise the barter system which men used before money was devised, and which even now obtains among certain simple peoples.

And next they raise a great outcry against other metals, as iron, than which they say nothing more pernicious could have been brought into the life of man. For it is employed in making swords, javelins, spears, pikes, arrows – weapons by which men are wounded, and which cause slaughter, robbery, and wars. . . .

It is claimed too, that lead is a pestilential and noxious metal, for men are punished by means of molten lead, as Horace describes in the ode addressed to the Goddess Fortune: 'Cruel Necessity ever goes before thee bearing in her brazen hand the spikes and wedges, while the awful hook and molten lead are also not lacking.'[1] In their desire to excite greater odium for this metal, they are not silent about the leaden balls of muskets, and they find in it the cause of wounds and death.

They contend that, inasmuch as Nature has concealed metals far within the depths of the earth, and because they are not necessary to human life, they are therefore despised and repudiated by the noblest, and should not be mined, and seeing that when brought to light they have always proved the cause of very great evils, it follows that mining is not useful to mankind, but on the contrary harmful and destructive. Several good men have been so perturbed by these tragedies that they conceive an intensely bitter hatred toward metals, and they wish absolutely that metals had never been created, or being created, that no one had ever dig them out. The more I commend the singular honesty, innocence, and goodness of such men, the more anxious shall I be to remove utterly and eradicate all error from their minds and to reveal the sound view, which is that the metals are most useful to mankind.

In the first place then, those who speak ill of the metals and refuse to make use of them, do not see that they accuse and condemn as wicked the Creator Himself, when they assert that He fashioned some things vainly and without good cause, and thus they regard Him as the Author of evils, which opinion is certainly not worthy of pious and sensible men.

In the next place, the earth does not conceal metals in her depths because she does not wish that men should dig them out, but because provident and sagacious Nature has appointed for each thing its place. She generates them in the veins, stringers, and seams in the rocks, as though in special vessels and receptacles for such material. The metals cannot be produced in the other elements because the materials for their formation are wanting. For if they were generated in the air, a thing that rarely happens, they could not find a firm resting-place, but by their own force and weight would settle down on to the ground. Seeing then that metals have their proper abiding place in the bowels of the earth, who does not see that these men do not reach their conclusions by good logic?

They say, 'Although metals are in the earth, each located in its own proper place where it originated, yet because they lie thus enclosed and hidden from sight, they should not be taken out.' But, in refutation of these attacks, which are so annoying, I will on behalf of the metals instance the fish, which we catch, hidden and concealed though they be in the water, even in the sea. Indeed, it is far stranger that man, a terrestrial animal, should search the interior of the sea than the bowels of the earth. For as birds are born to fly freely through the air, so are fishes born to swim through the waters, while to other creatures Nature has given the earth that they might live in it, and particularly to man that he might cultivate it and draw out of its caverns metals and other mineral products.

Note

1   Horace, *Odes* I, 35, lines 17–20.

De Re Metallica (1556), trans. H. C. and L. H. Hoover, London: Mining Magazine (1912), 6–8, 10–12.

Francis Bacon

New Atlantis

God blesse thee, my Sonne; I will give thee the greatest Iewell I have: For I will impart unto thee, for the Love of God and Men, a Relation of the true State of Salomons House. Sonne, to make you know the true state of Salomons House, I will keepe this order. First I will set forth unto you the End of our Foundation. Secondly, the Preparations and Instruments we have for our Workes. Thirdly, the severall Employments and Functions wherto our Fellowes are assigned. And fourthly, the Ordinances and Rites which we observe.

The End of our Foundation is the Knowledge of Causes, and Secrett Motions of Things; And the Enlarging of the bounds of Humane Empire, to the Effecting of all Things possible.

The Preparations and Instruments are these. We have large and deepe Caves of severall Depths: The deepest are sunke 600 Fathome: And some of them are digged and made under great Hills and Mountaines: So that if you reckon together the Depth of the Hill, and the Depth of the Cave, they are (some of them) above three Miles deepe. For we finde, that the Depth of a Hill, and the Depth of a Cave from the Flat, is the same Thing; Both remote alike, from the Sunn and Heavens Beames, and from the Open Aire. These Caves we call the Lower Region; And we use them for all Coagulations, Indurations, Refrigerations, and Conservations of Bodies. We use them likewise for the Imitation of Naturall Mines; And the Producing also of New Artificiall Mettalls, by Compositions and Materialls which we use, and lay ther for many yeares. Wee use them also sometimes, (which may seeme strange,) for Curing of some Diseases, and for Prolongation of Life, in some Hermits that choose to live ther, well accommodated of all things necessarie, and indeed live very long; By whom also we learne many things.

We have Burialls in severall Earths, wher we put diverse Cements, as the Chineses doe their Porcellane. But we have them in greater Varietie, and some of them more fine. We have also great variety of Composts, and Soiles, for the Making of the Earth Fruitfull.

We have High Towers; The Highest about halfe a Mile in Heigth; And some of them likewise set upon High Mountaines: So that the Vantage of the Hill with the Tower, is in the highest of them three Miles at least. . . .

We have also Parks, and Enclosures of all Sorts, of Beasts, and Birds; which wee use not onely for View or Rarenesse, but likewise for Dissections, and Trialls; That thereby we may take light, what may be wrought upon the Body of Man. Wherin we finde many strange Effects; As Continuing Life in them, though diverse Parts, which you account Vitall, be perished, and taken forth; Resussitating of some that seeme Dead in Appearance; And the like. We try also all Poysons, and other Medicines upon them, as well of Chyrurgery, as Phisicke. By Art likewise, we make them Greater, or Taller, then their Kinde is; And contrary-wise Dwarfe them and stay their Grouth: We make them more Fruitfull, and Bearing then their Kind is; and contrary-wise Barren and not Generative.

Also we make them differ in Colour, Shape, Activity, many wayes. We finde Meanes to make Commixtures and Copulations of diverse Kindes; which have produced many New Kindes, and then not Barren, as the generall Opinion is. We make a Number of Kindes, of Serpents, Wormes, Flies, Fishes, of Putrefaction; Wherof some are advanced (in effect) to be Prefect Creatures, like Beastes or Birds; And have Sexes, and doe Propagate. Neither doe we this by chance, but wee know before hand, of what Matter and Commixture, what Kinde of those Creatures will arise.

We have also Particular Pooles, wher we make Trialls upon Fishes, as we have said before of Beasts, and Birds.

We have also Places for Breed and Generation of those Kindes of Wormes, and Flies, which are of Speciall Use; Such as are with you your Silkwormes, and Bees.

The New Atlantis, ed. G.C. Moore-Smith (1627), Cambridge: Cambridge University Press (1900), 34–6

Andrew Jackson

Humanity has often wept

Humanity has often wept over the fate of the aborigines of this country, and Philanthropy has been long busily employed in devising means to avert it, but its progress has never for a moment been arrested, and one by one have many powerful tribes disappeared from the earth. To follow to the tomb the last of his race and to tread on the graves of extinct nations excite melancholy reflections. But true philanthropy reconciles the mind to these vicissitudes as it does to the extinction of one generation to make room for another.... Philanthropy could not wish to see this continent restored to the condition in which it was found by our forefathers. What good man would prefer a country covered with forests and ranged by a few thousand savages to our extensive Republic, studded with cities, towns, and prosperous farms, embellished with all the improvements which art can devise or industry execute, occupied by more than 12,000,000 happy people, and filled with all the blessings of liberty, civilization, and religion?

The present policy of the Government is but a continuation of the same progressive change by a milder process....

'Second Annual Message' (1830), in J. D. Richardson, (ed.), *A Compilation of the Messages and Papers of the Presidents, 1789–1897*, Washington, D.C.: Richardson (1908), 2: 520–1.

James Fenimore Cooper

Notions of the Americans

Once for all, dear Waller, I wish you to understand that – a few peaceable and half-civilized remains of tribes, that have been permitted to reclaim small portions of land, excepted – an inhabitant of New-York is actually as far removed from a savage as an inhabitant of London. The former has to traverse many hundred leagues of territory to enjoy even the sight of an Indian, in a tolerably wild condition; and the latter may obtain a similar gratification at about the same expense of time and distance, by crossing the ocean to Labrador. A few degraded descendants of the ancient warlike possessors of this country are indeed seen wandering among the settlements, but the Indian must now be chiefly sought west of the Mississippi, to be found in any of his savage grandeur.

Cases do occur, beyond a doubt, in which luckless individuals are induced to make their settlement in some unpropitious spot where the current of emigration obstinately refuses to run. These subjects of an unfortunate speculation are left to struggle for years in a condition between rude civilization, and one approaching to that of the hunter, or to abandon their possessions, and to seek a happier section of the country. Nine times in ten, the latter course is adopted. But when this tide of emigration has set steadily towards any favoured point for a reasonable time, it is absurd to seek for any vestige of a barbarous life among the people. The emigrants carry with them (I now speak of those parts of the country I have seen) the wants, the habits, and the institutions, of an advanced state of society. The shop of the artisan is reared simultaneously with the rude dwelling of the farmer. The trunks of trees, piled on each other, serve for both for a few years, and then succeed dwellings of wood, in a taste, magnitude, and comfort, that are utterly unknown to men of similar means in any other quarter of the world, which it has yet been my lot to visit. The little school-house is shortly erected at some convenient point, and a tavern, a store, (the American term for a shop of all sales,) with a few tenements occupied by mechanics, soon indicate the spot for a church, and the site of the future village. From fifty or a hundred of these centres of exertion, spread swarms that in a few years shall convert mazes of dark forests into populous, wealthy, and industrious counties. The manufactures of Europe, of the Indies, and of China, are seen exposed for sale, by the side of the coarse products of the country; and the same individual who vends the axe to fell the adjoining forest, can lay before your eyes a very tolerable specimen of Lyons silk, of English broadcloth, of Nankins, of teas, of coffees, or indeed of most of the more common luxuries of life. The number and quality of the latter increase with the growth of the establishment; and it is not too much to say, that an American village store, in a thriving part of the country, where the settlements are of twenty years' standing, can commonly supply as good an assortment of the manufactures of Europe, as a collection of shops in any European country town; and, if the general nature of their stock be considered, embracing, as it does, some of the products of all countries, one much greater.

As to wild beasts, savages, &c. &c. &c., they have no existence in these regions. A solitary bear, or panther, or even a wolf, wandering near the flocks of a country twenty years old, has an effect like that produced by an invasion. In the earlier days of the settlement, it is a task to chase the ravenous beasts from the neighbourhood. A price is offered for their heads, and for a time a mutual destruction against the flocks on one side, and the beasts on the other, is the consequence. In a year or two, this task is reduced to an occasional duty. In a few more, it is sought as an amusement: and ere the twenty years expire, the appearance of a wolf among the American farms is far less common than on the most ancient plains of certain parts of France. Every man has his rifle or his musket; and every man not only knows how, but he is fond of using them against such foes. Thus, you see, though wild beasts may be permitted, like Raphael's Seraphim, to encircle your pictures of American manners in faint relief, they must rarely indeed be permitted to enter into the action of the piece; more especially if the scene be laid in any of the settled portions of the three States that form the subject of this letter.

Notions of the Americans (1828), New York: Ungar (1963), 1: 245–7.

Edward Bellamy

Looking backward

Living as we do in the closing year of the twentieth century enjoying the blessings of a social order at once so simple and logical that it seems but the triumph of common sense, it is no doubt difficult for those whose studies have not been largely historical to realise that the present organization of society is, in its completeness, less than a century old. No historical fact is, however, better established than that till nearly the end of the nineteenth century it was the general belief that the ancient industrial system, with all its shocking social consequences, was destined to last, with possibly a little patching, to the end of time. How strange and wellnigh incredible does it seem that so prodigious a moral and material transformation as has taken place since then could have been accomplished in so brief an interval! … The almost universal theme of the writers and orators who have celebrated this bi-millennial epoch has been the future rather than the past, not the advance that has been made, but the progress that shall be made, ever onward and upward, till the race shall achieve its ineffable destiny. This is well, wholly well, but it seems to me that nowhere can we find more solid ground for daring anticipations of human development during the next one thousand years, than by 'Looking Backward' upon the progress of the last one hundred.

That this volume may be so fortunate as to find readers whose interest in the subject shall incline them to overlook the deficiencies of the treatment is the hope in which the author steps aside and leaves Mr. Julian West to speak for himself. . . .

… The industry and commerce of the country, ceasing to be conducted by a set of irresponsible corporations and syndicates of private persons at their caprice and for their profit, were intrusted to a single syndicate representing the people, to be conducted in the common interest for the common profit. The nation, that is to say, organized as the one great business corporation in which all other corporations were absorbed; it became the one capitalist in the place of all other capitalists, the sole employer, the final monopoly in which all previous and lesser monopolies were swallowed up, a monopoly in the profits and economies of which all citizens shared. The epoch of trusts had ended in The Great Trust. . . .

. . . The most violent foes of the great private monopolies were now forced to recognize how invaluable and indispensable had been their office in educating the people up to the point of assuming control of their own business. Fifty years before, the consolidation of the industries of the country under national control would have seemed a vary daring experiment to the most sanguine. But by a series of object lessons, seen and studied by all men, the great corporations had taught the people an entirely new set of ideas on this subject. They had seen for many years syndicates handling revenues greater than those of states, and directing the labours of hundreds of thousands of men with an efficiency and economy unattainable in smaller operations. It had come to be recognized as an axiom that the larger the business the simpler the principles that can be applied to it; that, as the machine is truer than the hand, so the system, which in a great concern does the work of the master's eye in a small business, turns out more accurate results. . . .

'Here we are at the store of our ward,' said Edith, as we turned in at the great portal of one of the magnificent public buildings I had observed in my morning walk. There was nothing in the exterior aspect of the edifice to suggest a store to a representative of the nineteenth century. There was no display of goods in the great windows, or any device to advertise wares, or attract custom. Nor was there any sort of sign or legend on the front of the building to indicate the character of the business carried on there; but instead, above the portal, standing out from the front of the building, a majestic life-sized group of statuary, the central figure of which was a female ideal of Plenty, with her cornucopia. Judging from the composition of the throng passing in and out, about the same proportion of the sexes among shoppers prevailed as in the nineteenth century. As we entered, Edith said that there was one of these great distributing establishments in each ward of the city, so that no residence was more than five or ten minutes' walk from one of them. It was the first interior of a twentieth-century public building that I had ever beheld, and the spectacle naturally impressed me deeply. I was in a vast hall full of light, received not only from the windows on all sides, but from the dome, the point of which was a hundred feet above. Beneath it, in the centre of the hall, a magnificent fountain played, cooling the atmosphere to a delicious freshness with its spray. The walls and ceiling were frescoed in mellow tints, calculated to soften without absorbing the light which flooded the interior. Around the fountain was a space occupied with chairs and sofas, on which many persons were seated conversing. Legends on the walls all about the hall indicated to what classes of commodities the counters below were devoted. Edith directed her steps towards one of these, where samples of muslin of a bewildering variety were displayed, and proceeded to inspect them.

Looking Backward (1889), London: Routledge (1922), 5, 6–7, 44, 45, 76–7.

Beatrice and Sidney Webb

Soviet communism: a new civilization

Governments in the past have seldom thought of deliberately changing the environment of their peoples. This is not explicitly set out, even in the twentieth-century textbooks of political science of the Western world, as one of the purposes of government. Yet how can mankind be improved, or even in any way changed, without changing its environment? The Soviet government naturally gives a large place, in its policy of the Remaking of Man, to measures for the transformation of the environment, alike of the dwellers in cities and of those in rural areas. Under this head come a whole series of colossal projects, many of them already being partially put in operation year by year, as opportunity permits. These range from gigantic schemes of artificial irrigation in order to keep back the inroads of the desert on the cultivated land, on the one hand; and of subsoil drainage of the huge part now made up of swamps and marshes, on the other, up to plans for an all-pervading electrification of the whole area of the USSR, and for the completion of a continuous network of roads and navigable waterways throughout the vast plain. We have perforce to confine ourselves here to the one important part of the environment constituted by the buildings, in and about which the 170 millions of people in the USSR spend so many hours out of the twenty-four; together with the various common services made necessary by the aggregation of these buildings, and of those who frequent them, in the multitude of villages, and notably in the rapid expansion of populous cities.

Soviet Communism: a new Civilisation, London: Longman (1941), 2: 928–9.

CHAPTER 5

Putting the Earth first

E. Conze, *The Bodhisattva and the hungry tigress*
Ovid, *The man from Samos*
W. Blake, *Auguries of innocence*
P. B. Shelley, *A vindication of natural diet*
H. Salt, *Animals' rights*
J. Muir, *The wild parks*
A. Leopold, *The land ethic*

Introduction

> Honourable representatives of the great saurians of an older creation, may you long enjoy your lilies and rushes, and be blessed now and then with a mouthful of terror-stricken man by way of dainty.
>
> (John Muir on alligators, quoted in Fox 1981)

Fundamental to Green thinking are the linked concepts of deep ecology and animal liberation. Proponents of such ideas argue that other species and indeed 'All life has intrinsic value' (Bunyard and Morgan-Grenville 1987: 281). Many believe that Greens can be separated from mere environmentalists by virtue of their adherence to such 'bio-ethics' (Dobson 1990: 48). Where others wish to conserve the whale, so as to maintain the delicate balance of an ocean eco-system for the benefit of the human race, the deep ecologist/ animal liberationist highlights the innate and absolute right of the whale to life. With the publication of Singer's *Animal Liberation* in 1976, a militant and growing animal rights movement, opposed to the extinction of whales and other species, and active against hunting and vivisection, gained a significant boost. Their tactics of direct action, including the destruction of property, have since been borrowed by the militant Earth First! movement of the 1980s and '90s. Earth First!, starting in the United States and spreading to Australia, Canada and the UK, has stressed an attitude that places the Earth and its species first and opposes all human-centred assaults on nature.

Sympathies for animal liberation and deep ecology run deep and have a long history; hunter–gatherers worshipped nature spirits, some even treated their animal prey with considerable respect. Vegetarianism has roots both in Eastern religion and amongst classical Greek philosophers. In our first extract we find a traditional Buddhist legend that goes even further than abstention from flesh, with the hero sacrificing himself to feed the hungry tigress and save her cubs. This act propels him to full enlightenment and Buddhahood. The tale captures the outlook not just of Buddhism but of other Eastern religions such as Jainism, Hinduism and Taoism. Although all contain elements of a deep ecology outlook, expressed in compassion for other species, they equally attack nature as an illusion or fetter from which the spiritual soul must escape in order to gain enlightenment. Taoism, though, affirms life, looks to nature and rejects the holy nihilism that Buddhism and Jainism may be accused of advocating. Although stronger and older in the East, a vegetarian tradition can also be found in the West. The Pythagorian school of mysticism refused to take life and looked back to the Golden Age when harmony ruled between species. It is an open question whether the mathematician himself was a vegetarian or whether, as with other aspects of the Pythagorian canon, vegetarianism was merely projected back on to him by later followers. Here we find the Roman author Ovid's (43 B.C.–A.D. 17) account of his life-affirming view.

William Blake (1757–1827), the English poet, holistic philosopher and inspiration of much twentieth-century Green thought is our next source. He was both a mystic, who looked to Boehme, Gnosticism and the Neoplatonists, and a child of the Enlightenment, an era that stressed human rights. His sentiments that:

> A robin redbreast in a cage
> Puts all Heaven in a rage

capture an ecological appreciation of nature. That we are all interconnected and what befalls one part of nature will influence the rest is lyrically captured by the poet. Blake's contemporary and fellow poet Shelley (1792–1822) held similar opinions. His appreciation of animal rights seems to have stemmed from his concern for human rights, equally voiced by other radical Romantics of his day. Inspired by the French revolution of 1789, his mother-in-law, Mary Wollstonecraft, had already extended the notions of 'liberté, fraternité, égalité' in *A Vindication of the Rights of Woman*. There was a satirical response on the rights of 'brutes' penned by an anti-feminist which may have influenced Shelley to look seriously at animal liberation. Shelley, as we shall see later, combined vegetarianism with a concern for social justice and issues of health. Vegetarianism, during this period, was closely linked with radical politics, prominent non-carnivores, including John Oswald, author of *The Cry of Nature*, fighting for the French republic (Thomas 1984: 296). Seventy years later we find an even stronger animal rights movement integrated with popular political radicalism. Henry Salt, an ex-Eton classics teacher and leading Green socialist of the 1880s and '90s is our next contributor. As well as an autobiography, vividly entitled *Seventy Years among Savages*, he produced biographies of Shelley and Thoreau, seeking to continue the link between human and animal liberation. 'On this point,' argued Salt,

> some of Shelley's detractors have done him more justice than his admirers; for the former have at least been consistent in arguing that his vegetarian proclivities were all of a piece with his 'pernicious' views on social and religious subjects, and with his utopian' belief in the ultimate perfectibility of man.

Salt's study *Animals' Rights* (from which the next passage is taken) examines many modern themes of the animal liberation movement, including hunting, vivisection and zoos.

Deep ecology as described by Aldo Leopold (1887–1949) and John Muir (1838–1914) cares for plants and the soil; reverence is extended from animals to all life and all that supports life. Yet vegetarianism may be rejected by belief in an eco-system, where predators must kill their prey to maintain a balance. Muir 'as an adult seldom ate meat' but was almost willing to offer his body to the alligator, like the would-be Buddha to the tigress. Leopold was a keen hunter but at least gave up the gun for a bow, a strategy that made his hunting so ineffective as to be almost humane. Writing in 1901, Muir claimed that wilderness heals humanity and illustrates how it was under assault. In the 1940s his fellow American Leopold outlined a 'land ethic' that sought to liberate the soil from the vivisection of overdevelopment. Deep ecology, a phrase coined by the Norwegian philosopher Naess, of course extends beyond the twentieth century, and can be found amongst hunter–gatherer groups, worshippers of Gaia and many animal liberationists; our first extract perhaps captures its essence with the greatest poetry (Naess 1973: 95).

Edward Conze

The Bodhisattva and the hungry tigress

The Buddha told the following story to Ananda: 'Once upon a time, in the remote past, there lived a king, Maharatha by name. He was rich in gold, grain, and chariots, and his power, strength, and courage were irresistible. He had three sons who were like young gods to look at. They were named Mahapranada, Mahadeva, and Mahasattva.

One day the king went for relaxation into a park. The princes, delighted with the beauties of the park and the flowers which could be seen everywhere, walked about here and there until they came to a large thicket of bamboos. There they dismissed their servants, in order to rest for a while. But Mahapranada said to his two brothers, 'I feel rather afraid here. There might easily be some wild beasts about, and they might do us harm.' Mahadeva replied, 'I also feel ill at ease, though it is not my body I fear for. It is the thought of separation from those I love which terrifies me.' Finally, Mahasattva said:

'No fear feel I, nor any sorrow either,
In this wide, lonesome wood, so dear to Sages.
My heart is filled with bursting joy,
For soon I'll win the highest boon.'

As the princes strolled about in the solitary thicket they saw a tigress, surrounded by five cubs, seven days old. Hunger and thirst had exhausted the tigress, and her body was quite weak. On seeing her, Mahapranda called out, 'The poor animal suffers from having given birth to the seven cubs only a week ago! If she finds nothing to eat, she will either eat her own young, or die from hunger!' Mahasattva replied, 'How can this poor exhausted creature find food? Mahapranada said, 'Tigers live on fresh meat and warm blood.' Mehadeva said, 'She is quite exhausted, overcome by hunger and thirst, scarcely alive and very weak. In this state she cannot possibly catch any prey. And who would sacrifice himself to preserve her life?' Mahaprana said, 'Yes, self-sacrifice is so difficult!' Mahasattva replied, 'It is difficult for people like us, who are so fond of our lives and bodies, and who have so little intelligence. It is not difficult, however, for others, who are true men, intent on benefiting their fellow creatures, and who long to sacrifice themselves. Holy men are born of pity and compassion. Whatever the bodies they may get, in heaven or on earth, a hundred times will they undo them, joyful in their hearts, so that the lives of others may be saved.'

Greatly agitated, the three brothers carefully watched the tigress for some time, and then went towards her. But Mahasattva thought to himself, 'Now the time has come for me to sacrifice myself! For a long time I have served this putrid body and given it beds and clothes, food and drink, and conveyances of all kinds. Yet it is doomed to perish and fall down, and in the end it will break up and be destroyed. How much better to leave this ungrateful body of one's own accord in good time! It cannot subsist for ever, because it is like urine which must come out. Today I will use it for a sublime deed. Then it will act for me as a boat which helps me to cross the ocean of birth and death. When I have renounced this futile body, a mere ulcer, tied to countless becomings, burdened with urine and excrement, unsubstantial like foam, full of hundreds of parasites – then I shall win the perfectly pure Dharma-body, endowed with hundreds of virtues, full of such qualities as trance and wisdom, immaculate, free form all substrate, changeless and without sorrow.' So, his heart filled with boundless compassion, Mahasattva asked his brothers to leave him alone for a while, went to the lair of the tigress, hung his cloak on a bamboo, and made the following vow:

'For the weal of the world I wish to win enlightenment, incomparably wonderful. From deep compassion I now give away my body, so hard to quit, unshaken in my mind. That enlightenment I shall now gain, in which nothing hurts and nothing harms, and which the Jina's sons have praised. Thus I shall cross to the Beyond of the fearful ocean of becoming which fills the triple world!'

The friendly prince then threw himself down in front of the tigress. But she did nothing to him. The Bodhisattva noticed that she was too weak to move. As a merciful man he had taken no sword with him. He therefore cut his throat with a sharp piece of bamboo, and fell down near the tigress. She noticed the Bodhisattva's body all covered with blood, and in no time ate up all the flesh and blood, leaving only the bones.

'It was I, Ananda, who at that time and on that occasion was that Prince Mahasattva.'

Edward Conze, *Buddhist Scriptures*, London: Penguin (1973), 24–5.

Ovid

The man from Samos

There was a man here, a Samian by birth, but he had fled forth from Samos and its rulers, and through hatred of tyranny was living in voluntary exile. He, though the gods were far away in the heavenly regions, still approached them with his thought, and what Nature denied to his mortal vision he feasted on with his mind's eye. And when he had surveyed all things by reason and wakeful diligence, he would give out to the public ear the things worthy of their learning and would teach the crowds, which listened in wondering silence to his words, the beginnings of the great universe, the causes of things and what their nature is: What God is, whence come the snows, what is the origin of lightning, whether it is Jupiter or the winds that thunder from the riven clouds, what causes the earth to quake, by what law the stars perform their courses, and whatever else is hidden from men's knowledge. He was the first to decry the placing of animal food upon our tables. His lips, learned indeed but not believed in this, he was the first to open in such words as these:

'O mortals, do not pollute your bodies with a food so impious! You have the fruits of the earth, you have apples, bending down the branches with their weight, and grapes swelling to ripeness on the vines; you have also delicious herbs and vegetables which can be mellowed and softened by the help of fire. Nor are you without milk or honey, fragrant with the bloom of thyme. The earth, prodigal of her wealth, supplies you her kindly sustenance and offers you food without bloodshed and slaughter. Flesh is the wild beasts' wherewith they appease their hunger, and yet not all, since the horse, the sheep, and cattle live on grass; but those whose nature is savage and untamed, Armenian tigers, raging lions, bears and wolves, all these delight in bloody food. Oh, how criminal it is for flesh to be stored away in flesh, for one greedy body to grow fat with food gained from another, for one live creature to go on living through the destruction of another living thing! And so in the midst of the wealth of good which Earth, the best of mothers, has produced, it is your pleasure to chew the piteous flesh of slaughtered animals with your savage teeth, and thus to repeat the Cyclops' horrid manners! And you cannot, without destroying other life, appease the cravings of your greedy and insatiable maw!

'But that pristine age, which we have named the golden age, was blessed with the fruit of the trees and the herbs which the ground sends forth, nor did men defile their lips with blood. Then birds plied their wings in safety through the heaven, and the hare loitered all unafraid in the tilled fields, nor did its own guilelessness hang the fish upon the hook. All things were free from treacherous snares, fearing no guile and full of peace. But after someone, an ill exemplar, whoever he was, envied the food of lions, and thrust down flesh as food into his greedy stomach, he opened the way for crime. It may be that, in the first place, with the killing of wild beasts the steel was warmed and stained with blood. This would have been justified, and we admit that creatures which menace our own lives may be killed without impiety. But, while they might be killed, they should never have been eaten.

'Further impiety grew out of that, and it is thought that the sow was first condemned to death as a sacrificial victim because with her curved snout she had rooted up the planted seeds and cut off the season's promised crop. The goat is said to have been slain at the avenging altars because he had browsed the grape-vines. These two suffered because of their own offences! But, ye sheep, what did you ever do to merit death, a peaceful flock, born for man's service, who bring us sweet milk to drink in your full udders, who give us your wool for soft clothing, and who help more by your life than by your death? What have the oxen done, those faithful, guileless beasts, harmless and simple, born to a life of toil? Truly inconsiderate he and not worthy of the gift of grain who could take off the curved plow's heavy weight and in the next moment slay his husbandman; who with his axe could smite that neck which was worn with toil for him, by whose help he had so often renewed the stubborn soil and planted so many crops. Nor is it enough that we commit such infamy: they made the gods themselves partners of their crime and they affected to believe that the heavenly ones took pleasure in the blood of the toiling bullock! A victim without blemish and of perfect form (for beauty proves his bane), marked off with fillets and with gilded horns, is set before the altar, hears the priest's prayer, not knowing what it means, watches the barley-meal sprinkled between his horns, barley which he himself laboured to produce, and then, smitten to his death, he stains with his blood the knife which he has perchance already seen reflected in the clear pool. Straight-away they tear his entrails from his living breast, view them with care, and seek to find revealed in them the purposes of heaven. Thence (so great is man's lust for forbidden food!) do you dare thus to feed, O race of mortals! I pray you, do not do it, but turn your minds to these my words of warning, and when you take the flesh of slaughtered cattle in your mouths, know and realize that you are devouring your own fellow-labourers. . . .'

Metamorphoses, Leob Classical Library, London: Heinemann, (1921), 2: 369–75.

William Blake

Auguries of innocence

To see a World in a Grain of Sand,
And a Heaven in a Wild Flower,
Hold Infinity in the palm of your hand,
And Eternity in an hour.
A robin redbreast in a cage
Puts all Heaven in a rage.
A dove-house fill'd with doves and pigeons
Shudders hell thro' all its regions.
A dog starv'd at his master's gate
Predicts the ruin of the State.
A horse misus'd upon the road
Calls to Heaven for human blood.
Each outcry of the hunted hare
A fibre from the brain does tear.
A skylark wounded in the wing,
A cherubin does cease to sing.
The game-cock clipt and arm'd for fight
Does the rising sun affright.
Every wolf's and lion's howl
Raises from Hell a Human soul.
The wild deer, wandering here and there,
Keeps the Human soul from care.
The lamb misus'd breeds public strife,
And yet forgives the butcher's knife.
The bat that flits at close of eve
Has left the brain that won't believe.
The owl that calls upon the night
Speaks the unbeliever's fright.
He who shall hurt the little wren
Shall never be belov'd by men.
He who the ox to wrath has mov'd
Shall never be by woman lov'd.
The wanton boy that kills the fly
Shall feel the spider's enmity.
He who torments the chafer's sprite
Weaves a Bower in endless Night. . . .

The Poetical Works of William Blake, Oxford: Oxford University Press (1914), 171–2

Percy Bysshe Shelley

A vindication of natural diet

Let not too much, however, be expected from this system. The healthiest among us is not exempt from hereditary disease. The most symmetrical, athletic, and long-lived, is a being inexpressibly inferior to what he would have been, had not the unnatural habits of his ancestors accumulated for him a certain portion of malady and deformity. In the most perfect specimen of civilized man something is still found wanting by the physiological critic. Can a return to nature, then, instantaneously eradicate predispositions that have been slowly taking root in the silence of innumerable ages? Indubitably not. All that I contend for is, that from the moment of the relinquishing all unnatural habits, no new disease is generated; and that the predisposition to hereditary maladies gradually perishes for want of its accustomed supply. In cases of consumption, cancer, gout, asthma, and scrofula, such is the invariable tendency of a diet of vegetables and pure water.

Those who may be induced by these remarks to give the vegetable system a fair trial, should, in the first place, date the commencement of their practice from the moment of their conviction. All depends upon the breaking through a pernicious habit resolutely and at once. Dr Trotter[1] asserts that no drunkard was ever reformed by gradually relinquishing his dram. Animal flesh, in its effects on the human stomach, is analogous to a dram. It is similar in the kind, though differing in the degree, of its operation. The proselyte to a pure diet must be warned to expect a temporary diminution of muscular strength. The subtraction of a powerful stimulus will suffice to account for this event. But it is only temporary, and is succeeded by an equable capability for exertion far surpassing his former various and fluctuating strength. Above all, he will acquire an easiness of breathing, by which the same exertion is performed with a remarkable exemption from that painful and difficult panting now felt by almost every one after hastily climbing an ordinary mountain. He will be equally capable of bodily exertion or mental application after as before his simple meal. He will feel none of the narcotic effects of ordinary diet. Irritability, the direct consequence of exhausting stimuli, would yield to the power of natural and tranquil impulses. He will no longer pine under the lethargy of *ennui*, that unconquerable weariness of life, more dreaded than death itself. He will escape the epidemic madness that broods over its own injurious notions of the Deity, and 'realizes the hell that priests and beldams feign.' Every man forms, as it were, his god from his own character; to the divinity of one of simple habits, no offering would be more acceptable than the happiness of his creatures. He would be incapable of hating or persecuting others for the love of God. He will find, moreover, a system of simple diet to be a system of perfect epicurism. He will no longer be incessantly occupied in blunting and destroying those organs from which he expects his gratification.

The pleasures of taste to be derived from a dinner of potatoes, beans, peas, turnips, lettuces, with a dessert of apples, gooseberries, strawberries, currants, raspberries, and in winter, oranges, apples, and pears, is far greater than is supposed. Those who wait until they can eat this plain fare with the sauce of appetite will scarcely join with the hypocritical sensualist at a lord mayor's feast, who declaims against the pleasures of the table. Solomon kept a thousand concubines, and owned in despair that all was vanity. The man whose happiness is constituted by the society of one amiable woman would find some difficulty in sympathising with the disappointment of this venerable debauchee.

Note

1   See Trotter on 'The Nervous Temperament'.

A Vindication of Natural Diet, London: Vegetarian Society (1813), 23–5.

H. Salt

Animals' rights

To take a wild animal from its free natural state, full of abounding egoism and vitality, and to shut it up for the wretched remainder of its life in a cell where it has just space to turn round, and where it necessarily loses every distinctive feature of its character – this appears to me to be as downright a denial as could well be imagined of the theory of animals' rights. Nor is there very much force in the plea founded on the alleged scientific value of their zoological institutions, at any rate in the case of the wilder and less tractable animals, for it cannot be maintained that the establishment of wild-beast shows is in any way necessary for the advancement of human knowledge. For what do the good people see who go to the gardens on a half-holiday afternoon to poke their umbrellas at a blinking eagle-owl, or to throw dog-biscuits down the expansive throat of a hippopotamus? Not wild beasts or wild birds, certainly, for there never have been or can be such in the best of all possible menageries, but merely the outer semblances and simulacra of the denizens of forest and prairie – poor spiritless remnants of what were formerly wild animals. . . .

All that has been said of hunting and coursing is applicable also – in a less degree, perhaps, but on exactly the same principle – to the sports of shooting and fishing. It does not matter, so far as the question of animals' rights is concerned, whether you run your victims to death with a pack of yelping hounds, or shoot him with a gun, or drag him from his native waters by a hook; the point at issue is simply whether man is justified in inflicting any form of death or suffering on the lower races for his mere amusement and caprice. There can be little doubt what answer must be given to this question.

Animals' Rights connected in Relations to Social Progress, London: George Bell (1892), 50–1, 77

J. Muir

The wild parks

None of Nature's landscapes are ugly so long as they are wild; and much, we can say comfortingly, must always be in great part wild, particularly the sea and the sky, the floods of light from the stars, and the warm, unspoilable heart of the earth, infinitely beautiful, though only dimly visible to the eye of imagination. The geysers, too, spouting from the hot underworld; the sturdy, long-lasting glaciers on the mountains, obedient only to the sun; Yosemite domes and the tremendous grandeur of rocky cañons and mountains in general – these must always be wild, for men can change them and mar them hardly more than can the butterflies that hover above them. But the continent's outer beauty is fast passing away, especially the plant part of it, the most destructible and most universally charming of all.

Only thirty years ago, the great Central Valley of California, five hundred miles long and fifty miles wide, was one bed of golden and purple flowers. Now it is ploughed and pastured out of existence, gone forever – scarce a memory of it left in fence corners and along the bluffs of the streams. The gardens of the Sierra, also, and the noble forests in both reserved and unreserved portions are sadly hacked and trampled, notwithstanding the ruggedness of the topography – all excepting those of the parks guarded by a few soldiers. In the noblest forests of the world, the ground, once divinely beautiful, is desolate and repulsive, like a face ravaged by disease. This is true also of many other Pacific coast and Rocky Mountain valleys and forests. The same fate, sooner or later, is awaiting them all, unless awakening public opinion comes forward to stop it. Even the great deserts in Arizona, Nevada, Utah, and New Mexico, which offer so little to attract settlers, and which a few years ago pioneers were afraid of, as places of desolation and death, are now taken as pastures at the rate of one or two square miles per cow, and of course their plant treasures are passing away – the delicate abronias, phloxes, gilias, etc. Only a few of the bitter, thorny, unbitable shrubs are left, and the sturdy cactuses that defend themselves with bayonets and spears.

Most of the wild plant wealth of the East also has vanished, – gone into dusty history. Only vestiges of its glorious prairie and woodland wealth remain to bless humanity in boggy, rocky, unploughable places. Fortunately, some of these are purely wild, and go far to keep Nature's love visible. White water-lilies, with rootstocks deep and safe in mud, still send up every summer a milky way of starry, fragrant flowers around a thousand lakes, and many a tuft of wild grass waves its panicles on mossy rocks, beyond reach of trampling feet, in company with saxifrages, bluebells, and ferns. Even in the midst of farmers' fields, precious sphagnum bogs, too soft for the feet of cattle, are preserved with their charming plants unchanged – chiognes, Andromeda, Kalmia, Limaea, Arethusa, etc. Calypso borealis still hides in the arbor vitae swamps of Canada, and away to the southward there are a few unspoiled swamps, big ones, where miasma, snakes, and alligators, like guardian angels, defend their treasures and keep them pure as paradise. And beside a' that and a' that, the East is blessed with good winters and blossoming clouds that shed white flowers over all the land, covering every scar and making the saddest landscape divine at least once a year. . . .

Of the four national parks of the West, the Yellowstone is far the largest. It is a big, wholesome wilderness on the broad summit of the Rocky Mountains, favoured with abundance of rain and snow – a place of fountains where the greatest of the American rivers take their rise. The central portion is a densely forested and comparatively level volcanic plateau with an average elevation of about eight thousand feet above the sea, surrounded by an imposing host of mountains belonging to the subordinate Gallatin, Wind River, Teton, Absaroka, and Snowy ranges. Unnumbered lakes shine in it, united by a famous band of streams that rush up out of hot lava beds, or fall from the frosty peaks in channels rocky and bare, mossy and bosky, to the main rivers, singing cheerily on through every difficulty, cunningly dividing and finding their way east and west to the two far-off seas. . . .

Camp out among the grass and gentians of glacier meadows, in craggy garden nooks full of Nature's darlings. Climb the mountains and get their good tidings. Nature's peace will flow into you as sunshine flows into trees. The winds will blow their own freshness into you, and the storms their energy, while cares will drop off like autumn leaves. As age comes on, one source of enjoyment after another is closed, but Nature's sources never fail, like a generous host, she offers here brimming cups in endless variety, served in a grand hall, the sky its ceiling, the mountains its walls, decorated with glorious paintings and enlivened with bands of music ever playing. The petty discomforts that beset the awkward guest, the unskilled camper, are quickly forgotten, while all that is precious remains. Fears vanish as soon as one is fairly free in the wilderness.

Our National Parks, Boston, Mass.: Houghton Mifflin (1901), 4–7, 37, 56–8.

Aldo Leopold

The land ethic

There are some who can live without wild things and some who cannot. These essays are the delights and dilemmas of one who cannot.

Like winds and sunsets, wild things were taken for granted until progress began to do away with them. Now we face the question whether a still higher 'standard of living' is worth its cost in things natural, wild, and free. For us of the minority, the opportunity to see geese is more important than television, and the chance to find a pasque-flower is a right as inalienable as free speech.

These wild things, I admit, had little human value until mechanization assured us of a good breakfast, and until science disclosed the dram of where they come from and how they live. The whole conflict thus boils down to a question of degree. We of the minority see a law of diminishing returns in progress; our opponents do not . . .

Conservation is getting nowhere because it is incompatible with our Abrahamic concept of land. We abuse land because we regard it as a commodity belonging to us. When we see land as a community to which we belong, we may begin to use it with love and respect. There is no other way for land to survive the impact of mechanized man, nor for us to reap from it the ethical harvest it is capable, under science, of contributing to culture. . . .

That land is a community is the basic concept of ecology, but that land is to be loved and respected is an extension of ethics. That land yields a cultural harvest is a fact long known, but latterly often forgotten. . . .

Such a view of land and people is, of course, subject to the lures and distortions of personal experience and personal bias. But wherever the truth may lie, this much is crystal-clear; our bigger-and-better society is now like a hypochondriac, so obsessed with its own economic health as to have lost the capacity to remain healthy. The whole world is so greedy for more bathtubs that it has lost the stability necessary to build them, or even to turn off the tap. Nothing could be more salutary at this stage than a little healthy contempt for a plethora of material blessings.

Perhaps such a shift of values can be achieved by reappraising things unnatural, tame, and confined in terms of things natural, wild, and free. . . .

It is inconceivable to me that an ethical relation to land can exist without love, respect, and admiration for land, and a high regard for its value. By value, I of course mean something far broader than mere economic value; I mean value in the philosophical sense.

Perhaps the most serious obstacle impeding the evolution of a land ethic is the fact that our educational and economic system is headed away from, rather than toward, an intense consciousness of land. Your true modern is separated from the land by many middlemen, and by innumerable physical gadgets. He has no vital relation to it; to him it is the space between cities on which crops grow. Turn him loose for a day on the land, and if the spot does not happen to be a golf link or 'scenic' area, he is bored stiff. If crops could be raised by hydroponics instead of farming, it would suit him very well. Synthetic substitutes for wood, leather, wool, and other natural products suit him better than the originals. In short, land is something he has 'outgrown'.

Almost equally serious as an obstacle to a land ethic is the attitude of the farmer for whom the land is still an adversary, or a taskmaster that keeps him in slavery. Theoretically, the mechanisation of farming ought to cut the farmer's chains, but whether it really does is debatable.

One of the requisites for an ecological comprehension of land is an understanding of ecology, and this is by no means co-existensive with 'education'; in fact much higher education seems deliberately to avoid ecological concepts. An understanding of ecology does not necessarily originate in courses bearing ecological labels; it is quite as likely to be labelled geography, botany, agronomy, history, or economics. This is as it should be, but whatever the label, ecological training is scarce.

The case for a land ethic would appear hopeless but for the minority which is in obvious revolt against these 'modern' trends.

A Sand County Almanac, New York: Oxford University Press (1949), vii, viii–ix, 223–4.

CHAPTER 6

Gaia

St Augustine, Concerning the Earth
R. Graves, The White Goddess
C. P. Gilman, Maternal Panteism
D. H. Lawrence, Pan in America

Introduction

> For Aristotelians, Platonists, and Paracelsians alike in the sixteenth century the
> world was conceived to be alive – and at all levels. It is not unusual to read
> theoretical accounts of the impregnation of the earth by astral seeds and of the
> resultant growth of metals in veins. This process was considered by many to be
> comparable to the growth of a human fetus.
>
> (Debus 1978: 34)

The Gaia hypothesis that the Earth works as a single self-sustaining unit, even a living
being, possibly a conscious one, was a motif of the Green movement in the 1980s and
1990s. Conceived by the biologist James Lovelock in the late 1960s, the theory holds that
our planet reacts to any form of atmospheric change to restore the best balance for life via
sophisticated global feedback mechanisms. Gaia, named after the ancient Greek Earth
goddess, whose title forms the root of both 'geology' and 'geography', has mixed
implications for the Greens. If Gaia is alive, we should revere her and act with modest
intentions towards her, perhaps. Yet, as Lovelock (1979) implies, within broad limits,
whatever harm we do to Gaia, its mechanisms can be expected to counter our attacks, the
need for an effective response is thus diminished.

The concept of an Earth goddess is nearly universal and certainly ancient. Egyptian,
Greek, Indian and Jewish traditions, to name but a few, provide us with female Earth
deities. Plato believed the Earth to be a single organism, albeit genderless. All forms of
paganism incorporate the idea of a living nature that gives humanity birth and should be
worshipped. Paganism (the word is derived from the Roman term for a country dweller)
was suppressed in Europe with great ferocity by the Christians, although traces of its
traditions may have been preserved by the witches persecuted in the seventeenth century.
In our first extract, St Augustine of Hippo (354–430), in his early Christian polemic *The
City of God*, pours venom on the worshippers of the various Earth goddesses in his native
North Africa. His descriptions encompass something of both Lovelock's modern scientific
concept and prehistoric paganism; the goddess is both a living presence that maintains life
and is worshipped by enthusiastic followers. The next sources are derived via Robert
Graves's (1895–1985) *The White Goddess*, his study of a sister moon deity. Some confusion
between Gaia and other nature goddesses results. First we find an account from the
Roman author Apuleius (123–*c*. 170 BC) of *The Golden Ass*, a Faustian story of a sorcerer's
assistant who is accidentally turned into an ass. Eventually a long pilgrimage and prayer
to Isis release him back to human form. The second is a pagan prayer to the goddess from

a twelfth-century herbal, providing us with an authentic hint of the hidden tradition of Earth goddess worship running through medieval Christendom. Whether the witches did indeed worship a female deity, and whether that deity was Gaia rather than the huntress Diana, is still an open question. Possible links between this Gaian tradition and pantheistic elements amongst Christian heretics is another interesting but largely unexplored area (see the pantheistic Ranter passage in Chapter Fifteen).

In 1915 we find the writer Charlotte Perkins Gilman (1860–1935) recounting in her feminist utopian novel *Herland* the implications of a society made up exclusively of women who celebrate a nurturing Mother Goddess. Wars, hierarchy, competition and capitalism have all been swept away. A matriarchical political economy that maintains harmony with the Earth is underpinned by a female-orientated religion. Perkins Gilman, we should note, throughout her activist life played down the notion of sexual and maternal roles, urging women to become involved in economics and society.

Finally, D. H. Lawrence (1885–1930) describes the Green Man, or the 'Great God Pan', variously seen as consort, servant or would-be male usurper of her role. Pantheism, in Lawrence's hands, is the virile male principle that provides all life with energy and vitality. That Lawrence inspired both mainstream Greens like Aldous Huxley and right-wingers such as the antisemitic co-founder of the organic agriculture movement Rolf Gardiner raises the suspicion of some critics (Freeman 1955: 201).

St Augustine

Concerning the Earth

23. *Concerning the earth, which Varro affirms to be a goddess, because that soul of the world which he thinks to be God pervades also this lowest part of his body, and imparts to it a divine force.*

Surely the earth, which we see full of its own living creatures, is one; but for all that, it is but a mighty mass among the elements, and the lowest part of the world. Why, then, would they have it to be a goddess? Is it because it is fruitful? Why, then, are not men rather held to be gods, who render it fruitful by cultivating it; but though they plough it, do not adore it? But, say they, the part of the soul of the world which pervades it makes it a goddess. As if it were not a far more evident thing, nay, a thing which is not called in question, that there is a soul in man. And yet men are not held to be gods, but (a thing to be sadly lamented), with wonderful and pitiful delusion, are subjected to those who are not gods, and than whom they themselves are better, as the objects of deserved worship and adoration. And certainly the same Varro, in the book concerning the select gods, affirms that there are three grades of soul in universal nature. One which pervades all the living parts of the body, and has not sensation, but only the power of life – that principle which penetrates into the bones, nails, and hair. By this principle in the world trees are nourished, and grow without being possessed of sensation, and live in a manner peculiar to themselves. The second grade of soul is that in which there is sensation. This principle penetrates into the eyes, ears, nostrils, mouth, and the organs of sensation. The third grade of soul is the highest, and is called mind, where intelligence has its throne. This grade of soul no mortal creatures except man are possessed of. Now this part of the soul of the world, Varro says, is called God, and in us is called Genius. And the stones and earth in the world, which we see, and which are not pervaded by the power of sensation, are, as it were, the bones and nails of God. Again, the sun, moon, and stars, which we perceive, and by which He perceives, are His organs of perception. Moreover, the ether is His mind; and by the virtue which is in it, which penetrates into the stars, it also makes them gods; and because it penetrates through them into the earth, it makes it the goddess Tellus, whence again it enters and permeates the sea and ocean, making them the god Neptune.

24. *Concerning the surnames of Tellus and their significations, which, although they indicate many properties, ought not to have established the opinion that there is a corresponding number of gods.*
... Then he adds that, because they gave many names and surnames to mother Tellus, it came to be thought that these signified many gods. 'They think,' says he, 'that Tellus is Ops, because the earth is improved by labour; Mother, because it brings forth much; Great, because it brings forth seed; Proserpine, because fruits creep forth from it; Vesta, because it is invested with herbs. And thus,' says he, 'they not at all absurdly identify other goddesses with the earth.' If, then, it is one goddess (though, if the truth were consulted, it is not even that), why do they nevertheless separate it into many? Let there be many names of one goddess, and let there not be as many goddesses as there are names.

The City of God, ed. Marcus Dods, Edinburgh: Clark (1913), 286–7, 289–90.

Robert Graves

The White Goddess

The most comprehensive and inspired account of the Goddess in all ancient literature is contained in Apuleius's *Golden Ass*, where Lucius invokes her from the depth of misery and spiritual degradation and she appears in answer to his plea; incidentally it suggests that the Goddess was once worshipped at Moeltre in her triple capacity of white raiser, red reaper and dark winnower of grain. The translation is by William Adlington (1566):

> About the first watch of the night when as I had slept my first sleep, I awaked with sudden fear and saw the moon shining bright as when she is at the full and seeming as though she leaped out of the sea. Then I thought with myself that this was the most secret time, when that goddess had most puissance and force, considering that all human things be governed by her providence; and that not only all beasts private and tame, wild and savage, be made strong by the governance of her light and godhead, but also things inanimate and

without life; and I considered that all bodies in the heavens, the earth, and the seas be by her increasing motions increased, and by her diminishing motions diminished: then as weary of all my cruel fortune and calamity, I found good hope and sovereign remedy, though it were very late, to be delivered from my misery, by invocation and prayer to the excellent beauty of this powerful goddess. Wherefore, shaking off my drowsy sleep, I arose with a joyful face, and, moved by a great affection to purify myself, I plunged my head seven times into the water of the sea; which number seven is convenable and agreeable to holy and divine things, as the worthy and sage philosopher Pythagoras hath declared. Then very lively and joyfully, though with a weeping countenance, I made this oration to the puissant goddess.

O'blessed Queen of Heaven, whether thou be the Dame Ceres which art the original and motherly source of all fruitful things on the earth, who after the finding of thy daughter Proserpine, through the great joy which thou didst presently conceive, didst utterly take away and abolish the food of them of old time, the acorn, and madest the barren and unfruitful ground of Eleusis to be ploughed and sown, and now givest men a more better and milder food; or whether thou be the celestial Venus, who, at the beginning of the world, didst couple together male and female with an engendered love, and didst so make an eternal propagation of human kind, being now worshipped within the temples of the Isle Paphos; or whether thou be the sister of the God Phoebus, who hast saved so many people by lightening and lessening with thy medicines the pangs of travail and art now adored at the sacred places of Ephesus; or whether thou be called terrible Proserpine by reason of the deadly howlings which thou yieldest, that hast power with triple face to stop and put away the invasion of hags and ghosts which appear unto men, and to keep them down in the closures of the Earth, which dost wander in sundry groves and art worshipped in divers manners; thou, which dost illuminate all the cities of the earth by thy feminine light; thou, which nourishest all the seeds of the world by thy damp heat, giving thy changing light according to the wanderings, near or far, of the sun: by whatsoever name of fashion or shape it is lawful to call upon thee, I pray thee to end my great travail and misery and raise up my fallen hopes, and deliver me from the wretched fortune which so long time pursued me. Grant peace and rest, if it please thee, to my adversities, for I have endured enough labour and peril. . . .'

When I had ended this oration, discovering my plaints to the goddess, I fortuned to fall again asleep upon that same bed; and by and by (for mine eyes were but newly closed) appeared to me from the midst of the sea a divine and venerable face, worshipped even of the gods themselves. Then, little by little, I seemed to see the whole figure of her body, bright and mounting out of the sea and standing before me: wherefore I purpose to describe her divine semblance, if the poverty of my human speech will suffer me, or the divine power give me a power of eloquence rich enough to express it. First, she had a great abundance of hair, flowing and curling, dispersed and scattered about her divine neck; on the crown on her head she bare many garlands interlaced with flowers, and in the middle of her forehead was a plain circlet in fashion of a mirror, or rather resembling the moon by the light it gave forth; and this was borne up on either side by serpents that seemed to rise from the furrows of the earth, and above it were blades of corn set out. Her vestment was of finest linen yielding diverse colours, somewhere white and shining, somewhere yellow like the crocus flower, somewhere rosy red, somewhere flaming; and (which troubled my sight and spirit sore) her cloak was utterly dark and obscure covered with shining black, and being wrapped round her from under her left arm to her right shoulder in manner of a shield, part of it fell down, pleated in most subtle fashion, to the skirts of her garment so that the welts appeared comely. Here and there upon the edge thereof and throughout its surface the stars glimpsed, and in the middle of them was placed the moon in mid-month, which shone like a flame of fire; and round about the whole length of the border of that goodly robe was a crown or garland wreathing unbroken, made with all flowers and all fruits. Things quite diverse did she bear: for in her right hand she had a timbrel of brass [*sistrum*], a flat piece of metal carved in manner of a girdle, wherein passed not many rods through the periphery of it; and when with her arm she moved these triple chords, they gave forth a shrill and clear sound. In her left hand she bare a cup of gold like unto a boat, upon the handle whereof, in the upper part which is best seen, an asp lifted up his head with a wide-swelling throat. Her odoriferous feet were covered with shoes interlaced and wrought with victorious palm. Thus the divine shape, breathing out the pleasant spice of fertile Arabia, disdained not with her holy voice to utter these words to me:

'Behold, Lucius, I am come; thy weeping and prayer hath moved me to succour thee. I am she that is the natural mother of all things, mistress and governess of all the elements, the initial progeny of worlds, chief of the powers divine, queen of all that are in Hell, the principal of them that dwell in Heaven, manifested alone and under one form of all the gods and goddesses [*deorum-dearum-que facies uniformis*]. At my will the planets of the sky, the wholesome winds of the seas, and the lamentable silences of hell be disposed; by name, my divinity is adored throughout the world, in divers manners, in variable customs, and by many names. For the Phrygians that are the first of all men call me. The Mother of the Gods at Pessinus; the Athenians, which are sprung from their own soil, Cecropian Minerva; the Cyprians, which are girt about by the sea, Paphian Venus; the Cretans which bear arrows, Dictynnian Diana; the Sicilians, which speak three tongues, Infernal Proserpine; the Eleusinians, their ancient goddess Ceres; some Juno, other Bellona, other Hecate, other Rhamnusia, and principally both sort of the Ethiopians which dwell in the Orient and are enlightened by the morning rays of the sun, and the Egyptians, which are excellent in all kind of ancient doctrine and by their proper cere-monies accustom to worship me, do call me by my true name, Queen Isis. Behold, I am come to take pity of thy fortune and turbulation; behold I am present to favour and aid thee; leave off thy weeping and lamentation, put away all thy sorrow, for behold the healthful day which is ordained by my providence.'

Much the same prayer is found in Latin in a twelfth-century English herbal (*Brit. Mus. MS. Harley*, 1585, *ff* 12*v*–13*r*):

> Earth, divine goddess, Mother Nature, who dost generate all things and bringest forth ever anew the sun which thou hast given to the nations; Guardian of sky and sea and of all Gods and powers; through thy influence all nature is hushed and sinks to sleep.... Again, when it pleases thee, thou sendest forth the glad daylight and nurturest life with thine eternal surety; and when the spirit of man passes, to thee it returns. Thou indeed art rightly named Great Mother of the Gods; Victory is in thy divine name. Thou art the source of the strength of peoples and gods; without thee nothing can either be born or made perfect; thou art mighty, Queen of the Gods. Goddess, I adore thee as divine, I invoke thy name; vouchsafe to grant that which I ask of thee, so shall I return thanks to thy godhead, with the faith that is thy due ....
>
> Now also I make intercession to you, all ye powers and herbs, and to your majesty: I beseech you, whom Earth the universal parent hath borne and given as a medicine of health to all peoples and hath put majesty upon, be now of the most benefit to humankind. This I pray and beseech you: be present here with your virtues, for she who created you hath herself undertaken that I may call you with the good will of him on whom the art of medicine was bestowed; therefore grant for health's sake good medicine by grace of these powers aforesaid ....

How the god of medicine was named in twelfth-century pagan England is difficult to determine; but he clearly stood in the same relation to the Goddess invoked in the prayers as Aesculapius originally stood to Athene, Thoth to Isis, Esmun to Ishtar, Diancecht to Brigit, Odin to Freya, and Bran to Danu.

The White Goddess (1948), London: Faber (1961), 70–3.

Charlotte Perkins Gilman

Maternal pantheism

As to Terry's criticism, it was true. These women, whose essential distinction of motherhood was the dominant note of their whole culture, were strikingly deficient in what we call 'femininity'. This led me very promptly to the conviction that those 'feminine charms' we are so fond of are not feminine at all, but mere reflected masculinity – developed to please us because they had to please us, and in no way essential to the real fulfillment of their great process. But Terry came to no such conclusion.

'Just you wait till I get out!' he muttered.

Then we both cautioned him. 'Look here, Terry, my boy! You be careful! They've been mighty good to us – but do you remember the anesthesia? If you do any mischief in this virgin land, beware of the vengeance of the Maiden Aunts! Come, be a man! It won't be forever.'

To return to the history:

They began at once to plan and build for their children, all the strength and intelligence of the whole of them devoted to that one thing. Each girl, of course, was reared in full knowledge of her Crowning Office, and they had, even then, very high ideas of the molding powers of the mother, as well as those of education.

Such high ideals as they had! Beauty, Health, Strength, Intellect, Goodness – for these they prayed and worked.

They had no enemies; they themselves were all sisters and friends. The land was fair before them, and a great future began to form itself in their minds.

The religion they had to begin with was much like that of old Greece – a number of gods and goddesses; but they lost all interest in deities of war and plunder, and gradually centered on their Mother Goddess altogether. Then, as they grew more intelligent, this had turned into a sort of Maternal Pantheism.

Here was Mother Earth, bearing fruit. All that they ate was fruit of motherhood, from seed or egg or their product. By motherhood they were born and by motherhood they lived – life was, to them, just the long cycle of motherhood.

But very early they recognized the need of improvement as well as of mere repetition, and devoted their combined intelligence to that problem – how to make the best kind of people. First this was merely the hope of bearing better ones, and then they recognized that however the children differed at birth, the real growth lay later – through education.

Then things began to hum.

As I learned more and more to appreciate what these women had accomplished, the less proud I was of what we, with all our manhood, had done.

You see, they had had no wars. They had had no kings, and no priests, and no aristocracies. They were sisters, and as they grew, they grew together – not by competition, but by united action.

We tried to put in a good word for competition, and they were keenly interested. Indeed, we soon found from their earnest questions of us that they were prepared to believe our world must be better than theirs. They were not sure; they wanted to know; but there was no such arrogance about them as might have been expected.

We rather spread ourselves, telling of the advantages of competition: how it developed fine qualities; that without it there would be 'no stimulus to industry'. Terry was very strong on that point.

'No stimulus to industry,' they repeated, with that puzzled look we had learned to know so well. '*Stimulus? To Industry?* But don't you *like* to work?'

'No man would work unless he had to,' Terry declared.

'Oh, no *man!* You mean that is one of your sex distinctions?'

'No, indeed!' he said hastily. 'No one, I mean, man or woman, would work without incentive. Competition is the – the motor power, you see.'

'It is not with us,' they explained gently, 'so it is hard for us to understand. Do you mean, for instance, that with you no mother would work for her children without the stimulus of competition?'

No, he admitted that he did not mean that. Mothers, he supposed, would of course work for their children in the home; but the world's work was different – that had to be done by men, and required the competitive element.

All our teachers were eagerly interested.

'We want so much to know – you have the whole world to tell us of, and we have only our little land! And there are two of you – the two sexes – to love and help one another. It must be a rich and wonderful world. Tell us – what is the work of the world, that men do – which we have not here?'

Herland (1915), London: Women's Press (1979), 58–60.

D. H. Lawrence

Pan in America

At the beginning of the Christian era, voices were heard off the coasts of Greece, out to sea, on the Mediterranean, wailing: 'Pan is dead! Great Pan is dead!'

The father of fauns and nymphs, satyrs and dryads and naiads was dead, with only the voices in the air to lament him. Humanity hardly noticed.

But who was he, really? Down the long lanes and overgrown ridings of history we catch odd glimpses of a lurking rustic god with a goat's white lightning in his eyes. A sort of fugitive, hidden among leaves, and laughing with the uncanny derision of one who feels himself defeated by something lesser than himself.

An outlaw, even in the early days of the gods. A sort of Ishmael among the bushes.

Yet always his lingering title: The Great God Pan. As if he was, or had been, the greatest.

Lurking among the leafy recesses, he was almost more demon than god. To be feared, not loved or approached. A man who should see Pan by daylight fell dead, as if blasted by lightning.

Yet you might dimly see him in the night, a dark body within the darkness. And then, it was a vision filling the limbs and the trunk of a man with power, as with new, strong-mounting sap. The Pan-power! You went on your way in the darkness secretly and subtly elated with blind energy, and you could cast a spell, by your mere presence, on women and on men. But particularly on women.

In the woods and the remote places ran the children of Pan, all the nymphs and fauns of the forest and the spring and the river and the rocks. These, too, it was dangerous to see by day. The man who looked up to see the white arms of a nymph flash as she darted behind the thick wild laurels away from him followed helplessly. He was a nympholept. Fascinated by the swift limbs and the wild, fresh sides of the nymph, he followed for ever, for ever, in the endless monotony of his desire. Unless came some wise being who could absolve him from the spell.

But the nymphs, running among the trees and curling to sleep under the bushes, made the myrtles blossom more gaily, and the spring bubble up with greater urge, and the birds splash with a strength of life. And the lithe flanks of the faun gave life to the oak groves, the vast trees hummed with energy. And the wheat sprouted like green rain returning out of the ground, in the little fields, and the vine hung its black drops in abundance, urging a secret.

Gradually men moved into cities. And they loved the display of people better than the display of a tree. They liked the glory they got of overpowering one another in war. And, above all, they loved the vainglory of their own words, the pomp of argument and the vanity of ideas.

So Pan became old and grey-bearded and goat-legged, and his passion was degraded with the lust of senility. His power to blast and to brighten dwindled. His nymphs became coarse and vulgar.

Till at last the old Pan died, and was turned into the devil of the Christians. The old god Pan became the Christian devil, with the cloven hoofs and the horns, the tail, and the laugh of derision. Old Nick, the Old Gentleman who is responsible for all our wicked-nesses, but especially our sensual excesses – this is all that is left of the Great God Pan.

It is strange. It is a most strange ending for a god with such a name. Pan! All! That which is everything has goat's feet and a tail! With a black face!

This really is curious.

Yet this was all that remained of Pan, except that he acquired brimstone and hell-fire, for many, many centuries. The nymphs turned into the nasty-smelling witches of a Walpurgis night, and the fauns that danced became sorcerers riding the air, or fairies no bigger than your thumb.

But Pan keeps on being reborn, in all kinds of strange shapes. There he was, at the Renaissance. And in the eighteenth century he had quite a vogue. He gave rise to an 'ism', and there were many pantheists, Wordsworth one of the first. They worshipped Nature in her sweet-and-pure aspect, her Lucy Gray aspect.

'Oft have I heard of Lucy Gray,' the school-child began to recite, on examination-day.

'So have I,' interrupted the bored inspector.

Lucy Gray, alas, was the form that William Wordsworth thought fit to give to the Great God Pan.

And then he crossed over to the young United States: I mean Pan did. Suddenly he gets a new name. He becomes the Oversoul, the Allness of everything. To this new Lucifer Gray of a Pan Whitman sings the famous *Song of Myself:* 'I am All, and All is Me.' That is: 'I am Pan, and Pan is me.'

The old goat-legged gentleman from Greece thoughtfully strokes his beard, and answers: 'All A is B, but all B is not A.' Aristotle did not live for nothing. All Walt is Pan, but all Pan is not Walt.

This, even to Whitman, is incontrovertible. So the new American pantheism collapses.

Then the poets dress up a few fauns and nymphs, to let them run riskily – oh, would there were any risk! – in their private 'grounds'. But, alas, these tame guinea-pigs soon became boring. Change the game.

We still *pretend* to believe that there is One mysterious Something-or-other back of Everything, ordaining all things for the ultimate good of humanity. It wasn't back of the Germans in 1914, of course, and whether it's back of the bolshevist is still a grave question. But still, it's back of *us*, so that's all right.

Alas, poor Pan! Is this what you've come to? Legless, hornless, faceless, even smileless, you are less than everything or anything, except a lie.

And yet here, in America, the oldest of all, old Pan is still alive. When Pan was greatest, he was not even Pan. He was nameless and unconceived, mentally. Just as a small baby new from the womb may say Mama! Dada! whereas in the womb it said nothing; so humanity, in the womb of Pan, said naught. But when humanity was born into a separate idea of itself, it said *Pan*.

In the days before man got too much separated off from the universe, he *was* Pan, along with all the rest.

As a tree still is. A strong-willed, powerful thing-in-itself, reaching up and reaching down. With a powerful will of its own it thrusts green hands and huge limbs at the light above, and sends huge legs and gripping toes down, down between the earth and rocks, to the earth's middle.

Here, on this little ranch under the Rocky Mountains, a big pine tree rises like a guardian spirit in front of the cabin where we live. Long, long ago the Indians blazed it. And the lightning, or the storm, has cut off its crest. Yet its column is always there, alive and changeless, alive and changing. The tree has its own aura of life. And in winter the snow slips off it, and in June it sprinkles down its little catkin-like pollen-tips, and it hisses in the wind, and it makes a silence within a silence. It is a great tree, under which the house is built. And the tree is still within the allness of Pan.

'Pan in America', *Phoenix*, New York: Viking (1936), 22–4.

CHAPTER 7

Philosophical holism

*J. al-din Rumi, The disagreement as to the
description and shape of the elephant*

W. Blake, To Thomas Butts

W. Blake, The marriage of Heaven and Hell

F. Engels, Natural dialectics

J. Smuts, The holistic doctrine of ecology

L. Mumford, The organic outlook

Introduction

Anarchist philosophers, pietist Christians, Taoist sages, somewhere, mixed up in the lot, a number of Romantic poets and nature mystics ... I grant this makes up an ungainly family tree. How many of these often cantankerous artists and thinkers could so much as share a conversation with one another? Yet I am sure they belong together. Not as an ideology, but as a sensibility – for the sacred, the organic, the personal.

(Roszak 1979: 139)

Green philosophy, as opposed to environmentalism, in its myriad forms always espouses holism, a method and outlook that examine the connections between things. Vitalism, the concept that living things are animated by a spark or force absent from the non-living is also sometimes embraced. Reductionist systems of thought that explore life by dividing it into tiny component parts are often criticized (Capra 1983; Merchant 1980). Deep ecology or 'eco-philosophy' is an ethical system (see Chapter Five) that places value upon other species and on nature, yet ethics of deep ecology have an epistemological basis in the science of ecology, which seeks to discover the links between species and investigate the interdependence of life.

Many oriental traditions have emphasized a holistic attitude and used natural metaphor to illustrate these beliefs; Taoism is an obvious example. Our first extract is the classic illustration of holistic principles drawn from a Sufi tradition. Sufism is the mystical 'heart' of Islam. Its early medieval wandering prophets and story tellers related paradoxical parables and preached universal love. Jalal al-din Rumi (1207–73), perhaps the greatest of these teachers, took earlier folk tales and retold them with greater flourish in his *Mathnavi*, a massive didactic work that ran to 26,000 verses. His message, a form of paradoxical Middle Eastern Zen, stressed that all is love and all life is linked by a process of transformation and interaction. The parable of the elephant states that if we look at only a part of something we will arrogantly misunderstand its nature.

Both Hegel and Goethe read and appreciated Rumi (Schmimmel 1987: 485). Goethe also read the Hindu sacred texts, as did Thoreau and Edward Carpenter, the British eco-socialist of the Victorian era. A Western tradition of holism also exists which can be traced back to the Neoplatonists of post-Roman Egypt, the Jewish cabbala, Henry More and Anne Conway as well as Boehme. It is perhaps both impossible and irrelevant to separate Western from Eastern streams of such thought. The influence of William Blake in synthesizing and transmitting such ancient knowledge cannot be understated. Inspired by

90

Blake, Ginsberg helped create the Beat movement of the 1950s; the Beats in turn provided the basis of the 1960s counter-culture out of which grew the modern Green movement. The Pre-Raphaelites, especially Dante Gabriel Rossetti (1828–82) – who popularized Blake's work – were strongly influenced by the poet. An examination of Green philosophical attitudes take us back not only to Eastern religion or, as Capra suggests, the New Physics of the twentieth century but almost always to Blake. Blake's equivalent of the elephant is his metaphor of 'The Marriage of Heaven and Hell', where balance can be created only through the dynamic interplay of opposites. His poem provides a mind-bending and dramatic rendition of an alternative means of perception. His many aphorisms are commonplaces of twentieth-century Green and mystical literature.

Engels may be seen as pedestrian in comparison with Blake and Rumi. He was certainly no poet, but along with his co-worker Marx believed in, and made use of, a holistic methodology. In this passage Engels traces the roots of a philosophy based on nature that stresses fluidity and movement from the ancient Heraclitus to Hegel. Ironically neither Engels, Hegel nor Marx drew 'Green' conclusions from their own clearly Green philosophy! As we have already seen and will continue to note, many thinkers have adhered to particular elements of a Green approach whilst rejecting the package as a whole. Thus, in contrast to the creators of Marxism, we find Jan Smuts (1879–1950), the noted general and Prime Minister of South Africa, stressing in 1935 the importance of ecology and holism. Author in 1926 of *Holism and Evolution*, unlike Marx and Engels he was clearly not just critical of environmental damage (they, after all, examine problems of pollution and soil erosion) but suspicious of unlimited industrial expansion. Yet as the ruler of a vast and exploited country he clearly cannot be described as a Green! Both Engels and Smuts enthusiastically argued that the study of the environment might provide 'iron laws of nature', a concept viewed as mechanistic and rigid by modern Green thinkers.

Lewis Mumford (1895–1989), the radical town planner and critic of the industrial 'mega-machine', can be seen as an unambiguously Green thinker. In books such as *The Culture of Cities* and *Technics and Civilisation* he criticized what he termed 'the mega-machine' of a society dominated by its industrial tools. According to Worster (1991: 320) he adhered to an 'organic ideology' inspired by his teacher, Patrick Geddes, and the school of ecological sociology at the University of Chicago in the 1920s. Using ecology, he built a holism that informed his social criticism and studies of city life. In turn he was read enthusiastically by post-war Green thinkers such as Bookchin and Commoner. He bridged the gap between ecology as science and ecological policy, living to see the fruit of this work in the growth of environmental awareness and the creation of Green parties in the 1970s and '80s. Here, like Engels, he examines the roots and form of an 'organic' outlook.

Jalal al-din Rumi

The disagreement as to the description and shape of the elephant

'The elephant was in a dark house: some Hindús had brought it for exhibition.

In order to see it, many people were going, every one, into that darkness.

As seeing it with the eye was impossible, (each one) was feeling it in the dark with the palm of his hand.

The hand of one fell on its trunk: he said, "This creature is like a water-pipe."

The hand of another touched its ear: to him it appeared to be like a fan.

Since another handled its leg, he said, "I found the elephant's shape to be like a pillar."

Another laid his hand on its back: he said, "Truly, this elephant was like a throne."

Similarly, whenever any one heard (a description of the elephant), he understood (it only in respect of) the part that he had touched.

On account of the (diverse) place (object) of view, their statements differed: one man entitled it '*dál*', another *alif*'.[1]

If there had been a candle in each one's hand, the difference would have gone out of their words.'

The eye of sense-perception is only like the palm of the hand: the palm hath not power to reach the whole of him (the elephant).

Note

1   I.e. crooked like the letter) or straight like the letter l.

Jaluddin Rumi, *The Mathnavi of Jaluddin Rumi*, Cambridge: Cambridge University Press (1930), IV: 71–2.

William Blake

To Thomas Butts, 22 November 1802

Now I a fourfold vision see,
And a fourfold vision is given to me;
'Tis fourfold in my supreme delight,
And threefold in soft Beulah's night,
And twofold always – May God us keep
From single vision and Newton's sleep!

William Blake

The Marriage of Heaven and Hell

The ancient Poets animated all sensible objects with Gods or Geniuses, calling them by the names and adorning them with the properties of woods, rivers, mountains, lakes, cities, nations, and whatever their enlarged and numerous senses could perceive.

And particularly they studied the Genius of each city and country, placing it under its Mental Deity;

Till a System was formed, which some took advantage of, and enslav'd the vulgar by attempting to realise or abstract the Mental Deities from their objects – thus began Priest-hood;

Choosing forms of worship from poetic tales.

And at length they pronounc'd that the Gods had order'd such things.

Thus men forgot that All Deities reside in the Human breast....

The ancient tradition that the world will be consumed in fire at the end of six thousand years is true, as I have heard from Hell.

For the cherub with his flaming sword is hereby commanded to leave his guard at tree of life; and when he does, the whole creation will be consumed and appear infinite and holy, whereas it now appears finite and corrupt.

This will come to pass by an improvement of sensual enjoyment.

But first the notion that man has a body distinct from his soul is to be expunged; this I shall do by printing in the infernal method, by corrosives, which in Hell are salutary and medicinal, melting apparent surfaces away, and displaying the infinite which was hid.

If the doors of perception were cleansed everything would appear to man as it is, infinite.

For man has closed himself up till he sees all things thro' narrow chinks of his cavern....

I was in a Printing-house in Hell, and saw the method in which knowledge is transmitted from generation to generation.

In the first chamber was a Dragon-Man, clearing away the rubbish from a cave's mouth; within, a number of Dragons were hollowing the cave.

In the second chamber was a Viper folding round the rock and the cave, and others adorning it with gold, silver, and precious stones.

In the third chamber was an Eagle with wings and feathers of air: he caused the inside of the cave to be infinite. Around were numbers of Eagle-like men who built palaces in the immense cliffs.

In the fourth chamber were Lions of flaming fire, raging around and melting the metals into living fluids.

In the fifth chamber were unnamed forms, which cast the metals into the expanse.

There they were received by Men who occupied the sixth chamber, and took the forms of books and were arranged in libraries.

The Giants who formed this world into its sensual existence, and now seem to live in it in chains, are in truth the causes of its life and the sources of all activity; but the chains are the cunning of weak and tame minds which have power to resist energy. According to the proverb, the weak in courage is strong in cunning.

Thus one portion of being is the Prolific, the other the Devouring. To the Devourer it seems as if the producer was in his chains; but it is not so, he only takes portions of existence and fancies that the whole.

But the Prolific would cease to be Prolific unless the Devourer, as a sea, received the excess of his delights.

Some will say: 'Is not God alone the Prolific?' I answer: 'God only Acts and Is, in existing beings or Men.

These two classes of men are always upon earth, and they should be enemies: whoever tries to reconcile them seeks to destroy existence.

Religion is an endeavour to reconcile the two.

Note. Jesus Christ did not wish to unite, but to separate them, as in the Parable of sheep and goats! And He says: 'I came not to send Peace, but a Sword.'

Messiah or Satan or Tempter was formerly thought to be one of the Antediluvians who are our Energies. . . .

An Angel came to me and said: 'O pitiable, foolish young man! O horrible! O dreadful state! Consider the hot, burning dungeon thou art preparing for thyself to all Eternity, to which thou art going in such career.'

I said: 'Perhaps you will be willing to show me my eternal lot, and we will contemplate together upon it, and see whether your lot or mine is most desirable.'

So he took me thro' a stable, and thro' a church, and down into the church vault, at the end of which was a mill. Thro' the mill we went, and came to a cave. Down the winding cavern we groped our tedious way, till a void boundless as a nether sky appear'd beneath us, and we held by the roots of trees, and hung over this immensity. But I said: 'If you please, we will commit ourselves to this void, and see whether Providence is here also. If you will not, I will.' But he answer'd: 'Do not presume, O young man, but as we here remain, behold thy lot which will soon appear when the darkness passes away.'

So I remain'd with him, sitting in the twisted root of an oak. He was suspended in a fungus, which hung with the head downward into the deep.

By degrees we beheld the infinite Abyss, fiery as the smoke of a burning city; beneath us, at an immense distance, was the sun, black but shining; round it were fiery tracks on which revolv'd vast spiders, crawling after their prey, which flew, or rather swum, in the infinite deep, in the most terrific shapes of animals sprung from corruption; and the air was full of them, and seem'd composed of them – these are Devils, and are called Powers of the Air. I now asked my companion which was my eternal lot? He said: 'Between the black and white spiders.'

But now, from between the black and white spiders, a cloud and fire burst and rolled thro' the deep, blackening all beneath; so that the nether deep grew black as a sea, and rolled with a terrible noise. Beneath us was nothing now to be seen but a black tempest, till looking East between the clouds and the waves we saw a cataract of blood mixed with fire, and not many stones' throw from us appear'd and sunk again the scaly fold of a monstrous serpent. At last, to the East, distant about three degrees, appear'd a fiery crest above the waves. Slowly it reared like a ridge of golden rocks, till we discover'd two globes of crimson fire, from which the sea fled away in clouds of smoke; and now we saw it was the head of Leviathan. His forehead was divided into streaks of green and purple like those on a tiger's forehead. Soon we saw his mouth and red gills hang just above the raging foam, tinging the black deep with beams of blood, advancing toward us with all the fury of a Spiritual Existence.

My friend the Angel climb'd up from his station into the mill: I remain'd alone, and then this appearance was no more; but I found myself sitting on a pleasant bank beside a river, by moonlight, hearing a harper, who sung to the harp; and his theme was: 'The man who never alters his opinion is like standing water, and breeds reptiles of the mind.'

But I arose and sought for the mill, and there I found my Angel, who, surprised, asked me how I escaped.

I answer'd: 'All that we saw was owing to your metaphysics; for when you ran away, I found myself on a bank by moonlight hearing a harper. But now we have seen my eternal lot, shall I show you yours?' He laugh'd at my proposal; but I, by force, suddenly caught him in my arms, and flew westerly thro' the night, till we were elevated above the earth's shadow; then I flung myself with him directly into the body of the sun. Here I clothed myself in white, and taking in my hand Swedenborg's volumes, sunk from the glorious clime, and passed all the planets till we came to Saturn. Here I stay'd to rest, and then leap'd into the void between Saturn and the fixed stars.

'Here,' said I, 'is your lot, in this space – if space it may be call'd.' Soon we saw the stable and the church, and I took him to the altar and open'd the Bible, and lo! it was a deep pit, into which I descended, driving the Angel before me. Soon we saw seven houses of brick. One we enter'd; in it were a number of monkeys, baboons, and all of that species, chain'd by the middle, grinning and snatching at one another, but withheld by the short-ness of their chains. However, I saw that they sometimes grew numerous, and then the weak were caught by the strong, and with a grinning aspect, first coupled with, and then devour'd, by plucking off first one limb and then another, till the body was left a helpless trunk. This, after grinning and kissing it with seeming fondness, they devour'd too; and here and there I saw one savourily picking the flesh off of his own tail. As the stench terribly annoy'd us both, we went into the mill, and I in my hand brought the skeleton of a body, which in the mill was Aristotle's Analytics.

So the Angel said: 'Thy phantasy has imposed upon me, and thou oughtest to be ashamed.'

I answer'd: 'We impose on one another, and it is but lost time to converse with you whose works are only Analytics.'

'The Marriage of Heaven and Hell' (1790), *The Poetical Works of William Blake*, London: Oxford University Press (1914), 189–90, 196–206.

Friedrich Engels

Natural dialectics

In the meantime, along with and after the French philosophy of the eighteenth century had arisen the new German philosophy, culminating in Hegel. Its greatest merit was the taking up again of dialectics as the highest form of reasoning. The old Greek philosophers were all born natural dialecticians, and Aristotle, the most encyclopaedic intellect of them, had already analysed the most essential forms of dialectic thought. The newer philosophy, on the other hand, although in it also dialectics had brilliant exponents (e.g. Descartes and Spinoza), had, especially through English influence, become more and more rigidly fixed in the so-called metaphysical mode of reasoning, by which also the French of the eighteenth century were almost wholly dominated, at all events in their special philosophical work. Outside philosophy in the restricted sense, the French nevertheless produced masterpieces of dialectics. We need only call to mind Diderot's 'Le Neveu de Rameau', and Rousseau's 'Discours sur l'origine et les fondements de l'inégalité parmi les hommes'. We give here, in brief, the essential character of these two modes of thought.

When we consider and reflect upon Nature at large or the history of mankind or our own intellectual activity, at first we see the picture of an endless entanglement of relations and reactions, permutations and combinations, in which nothing remains what, where and as it was, but everything moves, changes, comes into being and passes away. We see, therefore, at first the picture as a whole, with its individual parts still more or less kept in the background; we observe the movements, transitions, connections, rather than the things that move, combine and are connected. This primitive, naïve but intrinsically correct conception of the world is that of ancient Greek philosophy, and was first clearly formulated by Heraclitus; everything is and is not, for everything is fluid, is constantly changing, constantly coming into being and passing away.

But this conception, correctly as it expresses the general character of the picture of appearances as a whole, does not suffice to explain the details of which this picture is made up, and so long as we do not understand these, we have not a clear idea of the whole picture. In order to understand these details we must detach them from their natural or historical connection and examine each one separately, its nature, special causes, effects, etc. This is, primarily, the task of natural science and historical research: branches of science which the Greeks of classical times, on very good grounds, relegated to a subordinate position, because they had first of all to collect materials for these sciences to work upon. A certain amount of natural and historical material must be collected before there can be any critical analysis, comparison, and arrangement in classes, orders, and species. The foundations of the exact natural sciences were, therefore, first worked out by the Greeks of the Alexandrian period, and later on, in the Middle Ages, by the Arabs. Real natural science dates from the second half of the fifteenth century, and thence onward it had advanced with constantly increasing rapidity. The analysis of Nature into its individual parts, the grouping of the different natural processes and objects in definite classes, the study of the internal anatomy of organic bodies in their manifold forms – these were the fundamental conditions of the gigantic strides in our knowledge of Nature that have been made during the last four hundred years. But this method of work has also left us as legacy the habit of observing natural objects and processes in isolation, apart from their connection with the vast whole; of observing them in repose, not in motion; as constants, not as essentially variables; in their death, not in their life. And when this way of looking at things was transferred by Bacon and Locke from natural science to philosophy, it begot the narrow, metaphysical mode of thought peculiar to the last century.

To the metaphysician, things and their mental reflexes, ideas, are isolated, are to be considered one after the other and apart from each other, are objects of investigation fixed, rigid, given once for all. He thinks in absolutely irreconcilable antitheses. 'His communication is "yea, yea; nay, nay"; for whatsoever is more than these cometh of evil.' For him a thing either exists or does not exist; a thing cannot at the same time be itself and something else. Positive and negative absolutely exclude one another; cause and effect stand in a rigid antithesis one to the other.

At first sight this mode of thinking seems to us very luminous, because it is that of so-called sound common sense. Only sound common sense, respectable fellow that he is, in the homely realm of his own four walls, has very wonderful adventures directly he ventures out into the wide world of research. And the metaphysical mode of thought, justifiable and necessary as it is in a number of domains whose extent varies according to the nature of the particular object of investigation, sooner or later reaches a limit, beyond which it becomes one-sided, restricted, abstract, lost in insoluble contradictions. In the contemplation of individual things, it forgets the connection between them; in the contemplation of their existence, it forgets the beginning and end of that existence; of their repose, it forgets their motion. It cannot see the wood for the trees.

For everyday purposes we know and can say, e.g., whether an animal is alive or not. But, upon closer inquiry, we find that this is, in many cases, a very complex question, as the jurists know very well. They have cudgelled their brains in vain to discover a rational limit beyond which the killing of the child in its mother's womb is murder. It is just as impossible to determine absolutely the moment of death, for physiology proves that death is not an instantaneous, momentary phenomenon, but a very protracted process.

In like manner, every organic being is every moment the same and not the same; every moment it assimilates matter supplied from without, and gets rid of other matter; every moment some cells of its body die and others build themselves anew; in a longer or shorter time the matter of its body is completely renewed, and is replaced by other molecules of matter, so that every organic being is always itself, and yet something other than itself.

Further, we find upon closer investigation that the two poles of an antithesis, positive and negative, e.g., are as inseparable as they are opposed, and that despite all their opposition, they mutually interpenetrate. And we find, in like manner, that cause and effect are conceptions which only hold good in their application to individual cases; but as soon as we consider the individual cases in their general connection with the universe as a whole, they run into each other, and they become confounded when we contemplate that universal action and reaction in which causes and effects are eternally changing places, so that what is effect here and now will be cause there and then, and *vice versa*.

None of these processes and modes of thought enters into the framework of metaphysical reasoning. Dialectics, on the other hand, comprehends things and their representations, ideas, in their essential connection, concatenation, motion, origin, and ending. Such processes as those mentioned above are, therefore, so many corroborations of its own method of procedure.

Nature is the proof of dialectics, and it must be said for modern science that it has furnished this proof with very rich materials increasing daily, and thus has shown that, in the last resort, Nature works dialectically and not metaphysically; that she does not move in the eternal oneness of a perpetually recurring circle, but goes through a real historical evolution. In this connection Darwin must be named before all others. He dealt the metaphysical conception of Nature the heaviest blow by his proof that all organic beings, plants, animals, and man himself, are the products of a process of evolution going on through millions of years. But the naturalists who have learned to think dialectically are few and far between, and this conflict of the results of discovery with preconceived modes of thinking explains the endless confusion now reigning in theoretical natural science, the despair of teachers as well as learners, of authors and readers alike.

Socialism: Utopian and Scientific (1892), Moscow: Progress (1970), 49–53.

Jan Smuts

The holistic doctrine of ecology

Professor Bews has rightly stressed the necessity for the holistic viewpoint, of which the ecological method is a special scientific application. The world is not a chaos, a chance selection of items and fragments. It is a closely interwoven system of patterns. What we in our human way call plan and design is present everywhere. This is not to be understood as naïve anthropomorphism. Our most painstaking effort at understanding the world discloses certain dominant features in it – rhythm, regularity, inter-connexion, and linkings up, an interplay of active relationships which is creative of structures, forms, patterns. Such is reality – a vast Pattern of patterns. And to trace these patterns or wholes is to discover the lineaments of beauty in all its forms, whether we call them beauty or truth or good. They are all, but holistic harmonies in the nature of things. Nothing exists for itself alone; there are no isolated units, but only structured patterns and inter-relations, from the primordial electrons to the most developed physical or moral or social complexes in the universe.

This is the holistic doctrine which underlies ecology. The organism is not itself alone and in isolation. As a unit it is a mere static abstraction. The real dynamic unit is the organism functioning in its environment. This complex concept is the real biological unit and starting point. Life is living, and living is an active reciprocal relation between organism and environment. This is the central concept in ecology, and it has already led to a revolution of our biological sciences. It enables us not only to account for individual development, but also for the existence and development of communities, societies, their histories, phases, and climaxes and all the complexes which we find in the living nature. In the end they all follow certain large rhythms which prevail in nature. Plant communities, such as forests, grasslands, scrub bush, associations, and the like, are but ecological products of climate and other large or minor rhythms in nature. They form a moving equilibrium of patterns, expressive of something deeper in nature. The small patterns are thus grouped into the larger patterns, all expressive of the inner rhythmic nature of reality. The vision of this Harmony is what the gods feed on, and what mortals strive for, according to the Platonic myth.

In biology it is possible to trace the laws of ecology, because the environmental factors are regular, or at least definite and certain, and not arbitrary. Weather may change but it is an objective fact, which we can build on, and which can be statistically dealt with.

Not so in human affairs; and here comes the difficulty of applying ecological laws in the human sciences. In biology the environment is given, is objective fact, and its effects can be ascertained and formulated in laws. Man, however, shows his superiority to nature by largely creating his own environment. His intelligence enables him to circumvent nature, to command the forces of nature, and thus to alter the natural situation into what suits him. Thus the subjective human factor enters into his environment. The progress of scientific knowledge and technology is continuously revolutionizing our human environment. Under such circumstances, one could only apply the ecological laws, derived from biology, with caution and circumspection to human affairs. And this is not the only snag. The human organism is the most plastic and adaptable and undifferentiated of all. It is curious how this most complex of all organisms is, in many respects, the most primitive and unspecialized. Man therefore easily adapts himself to the most diverse and extreme conditions. While floras and faunas disappeared in the Great Ice Age, he found it only a greater spur to his progress. His innate adaptability and his intelligence have made him largely independent of environmental conditions and even of natural laws. No wonder that ecology will have a far more formidable task to perform in the human sciences than in scientific biology. Even here, ecology will supply scientific clues which in the end may lead us on to the right track. The adventure of science will lead us more and more towards the human; and ecology, and especially holism, may prove far more fruitful in that great quest than orthodox scientists may be prepared to admit today. At any rate, necessity is laid on us, and we dare not sit still in a world today fuller of dangers of our race than ever before. We must move on, and science appears to be the royal road of advance. We cannot accept at their face value the philosophical speculations of a Spengler – that civilization is an organism which grows and decays of its own inner unalterable laws, and that nothing can arrest the disappearance of our own civilization. Nor can we accept – in the face of all that genetics has taught us – that environment is all, and the organism merely a creature of it. Somewhere between lies the truth we are after, between the one-sided environmental and organismic views. Human Ecology, in doing justice to both the great factors that shape our fate, may yet prove the scientific reconciliation between these extreme views. My reflections on science, on the nature of knowledge, and on human affairs have only deepened my own impression that Holism in its various applications is the pathway to explore to the future of science and philosophy, and to human welfare.

'Introduction' to J, Bews, *Human Ecology*, Natal: Natal University Press (1935), x–xii.

Lewis Mumford

The organic outlook

The orientation of thought toward the realities of organic life, something that in the eighteenth century went no further than the intuitions of the poet or the naturalist, had by the end of the nineteenth century become so pervasive that it entered even into the hitherto lifeless realm of mechanics: the telephone, the phonograph, the motion picture, the airplane, sprang out of an interest in the functions of organisms and could not have developed without a scientific knowledge of their processes. Steadily, for the past generation, a transformation has been going on in every department of thought: a re-location of interest from mechanism to organism, a change from a world in which material bodies and mechanical motion alone were real to a world in which invisible rays and emanations, in which human projections and dreams, are as real as any immediately visible or external phenomenon – as real and on occasion more important.

Nineteenth century industry had been mainly concerned, in its paleotechnic phase, with the inorganic processes of factory, steel mill, and mine. The first significant revelation of experimental biological knowledge to industry came through Pasteur's researches on the diseases of silkworms, and the rôle of ferments in wines. That knowledge, which was to lay the foundation for modern hygiene and medicine, did not go into circulation until the eighteen-seventies. It needed the triumphant demonstrations of medicine in the ensuing decades to give authority to a new world-view, which accords to the organism and the world of life the priority that had been accorded, from the seventeenth century on, to the machine and to a universe whose cold mechanical perfection was described by physics and astronomy.

The leading ideas of this organic order may be briefly summed up.

First: the primacy of life, and of autonomous but perpetually inter-related organisms as vehicles of life. Each organism has its own line of growth, that of its species, its own curve of development, its own span of variations, its own pattern of existence. To maintain its life-shape the organism must constantly alter it and renew itself by entering into active relations with the rest of the environment. Even the most sessile and sleepy forms of life must seize energy in order to maintain their equilibrium: thus the organism changes, by no matter what infinitesimal amounts, the balance of the environment; and the failure to act and re-act means either the temporary suspension of life or its final end. Not merely is the organism implicated in its environment in space: it is also implicated in time, through the biological phenomena of inheritance and memory; and in human societies it is even more consciously implicated through the necessity of assimilating a complicated social heritage which forms, as it were, a second environment.

Human beings and groups are the outcomes of an historic complex, their inheritance, and they move toward a conditioned but uncertain destination, their future. The assimilation of the past and the making of the future are the two ever-present poles of existence in a human community. In so far as Aristotle appreciated the future, as potentiality and possibility, he was more truly a sociologist than those thinkers of the past century whose minds, even when dealing with society, have stopped short with time-categories that completely suffice only for elementary mechanics.

The autonomy of the organism, so characteristic of its growth, renewal, and repair, does not lead to isolation in either time or space. On the contrary, every living creature is part of the general web of life: only as life exists in all its processes and realities, from the action of the bacteria upward, can any particular unit of it continue to exist. As our knowledge of the organism has grown, the importance of the environment as a co-operative factor in its development has become clearer; and its bearing upon the development of human societies has become plainer, too. If there are favorable habitats and favorable forms of association for animals and plants, as ecology demonstrates, why not for men? If each particular natural environment has its own balance, is there not perhaps an equivalent of this in culture? Organisms, their functions, their environments: people, their occupations, their workplaces and living-places, form inter-related and definable wholes.

Such questions as yet can evoke only tentative answers; but they provide a new starting point for investigation. And from the negative processes, the destruction and deterioration of the environment through man's misuse, much has already been learned: Marsh's classic treatise on the Earth and Man was followed by the highly intelligent surveys of natural resources, in terms of potential human use, by Major Powell and Raphael Pumpelly, and the later conservationists, from Van Hise to MacKaye. Beginning with Kropotkin's Mutual Aid the study of human ecology has taken a more positive turn: witness Huntington's studies of civilization and climate, the urban investigations of the Chicago school of sociologists, and above all, Patrick Geddes's lifelong effort to develop a sociology on the basis of biology, and a social art on the positive foundation of our biological, psychological and sociological knowledge. In the doctrines of emergents, organisms, and wholes, particularly as set forth in Lloyd Morgan and Whitehead, lies the outline of an appropriate metaphysics.

In emphasizing the importance of this new orientation toward the living and the organic, I expressly rule out false biological analogies between societies and organisms: Herbert Spencer and others pushed these to the point of absurdity. Such analogies sometimes provide useful suggestions, suggestions no less practical than those derived – with equally little realism – from the machine. But the point is that our knowledge directs attention to parallel processes, parallel conditions and reactions; and it gives rise to related pictures of the natural and the cultural environments, considered as wholes, within which man finds his life and being and drama.

So long as the machine was uppermost, people thought quantitatively in terms of expansion, extension, progress, mechanical multiplication, power. With the organism uppermost we begin to think qualitatively in terms of growth, norms, shapes, inter-relationships, implications, associations, and societies. We realize that the aim of the social process is not to make men more powerful, but to make them more completely developed, more human, more capable of carrying on the specifically human attributes of culture – neither snarling carnivores nor insensate robots. Once established, the vital and social order must subsume the mechanical one, and dominate it: in practice as well as in thought. In social terms, this means a re-orientation not only from mechanism to organism, but from despotism to symbiotic association, from capitalism and fascism to co-operation and basic communism.

The Culture of Cities, London: Secker & Warburg (1938), 300–4.

CHAPTER 8

The web of life

Linnaeus, Simple, beautiful, and instructive

G. White, Letter XXXV

R. W. Emerson, Nature

C. Darwin, The complex relations of all animals and plants

P. Kropotkin, Mutual aid

Introduction

> So influential has their branch of science become that our time might well be
> called the 'Age of Ecology.'

(Worster 1991: xiii)

It is difficult to believe that a Green philosophy could have developed without a growing understanding of the natural world and the development of the science we now call ecology. Although the term, coined in 1869, is of recent origin, Worster in his history of the subject traces its prehistory back centuries earlier. He distinguished between 'arcadian' and 'imperial' forms of the subject. The arcadian ecologist, epitomized for him by Henry David Thoreau, believes that nature 'has an order, a pattern, that we humans are bound to understand and respect and preserve' (Worster 1991: ix). The imperialist, of whom Engels in *The Dialectics of Nature* is a splendid example, seeks to understand nature in order to control nature more efficiently for his species' exclusive benefit (Parsons 1977). The sciences of life have played a contradictory role, providing greater understanding and sympathy for the natural world yet equally increasing our ability to dominate, control and destroy.

Perhaps the most important ecologist prior to the invention of the term was Darwin, who showed, through his concept of evolution, our closeness to other animals. Ecology in particular and the environmental sciences in general have gradually shown how closely our species is linked to every other in a web or net of life connections. We have also become more aware of the damage done by our actions to our environment and ultimately to ourselves. The nineteenth-century geographers such as Somerville, Reclus and Marsh found themselves studying deforestation, extinction, pollution and soil erosion. Ecology strengthens the holistic philosophy of the Green movement, placing it on a rational rather than religious basis.

Ignoring ancient Chinese and classical Greek endeavours, the first systematic attempts to understand the environment came with the natural theologians of the seventeenth century. Religious writers sought to understand the beauty and efficiency of a world created by God's hand through what came to be known as 'the economy of nature'. Both ecology and economics derive from the Greek world *oikos*, which means 'house'. The pioneering Swedish naturalist Carl von Linné (1707–78) set about describing the 'furni-ture' of this house in a scheme still used by modern botanists and zoologists. His academic thesis, written in 1749, entitled *The Oeconomy of Nature*, has been claimed as 'the single most important summary of an ecological point of view still in its infancy' (Worster 1991:

33). Along with the natural theologians such as John Ray and Sir Kenelm Digby he argued that God provided for all life and maintained a balance between species.

Our second extract comes from the naturalist and clergyman Gilbert White (1720–93), who, continuing this tradition of studying nature for the benefit of God rather than Gaia, showed a keen interest in the environs of his parish of Selborne, Hampshire. Described as 'seminal' for 'the modern study of ecology', *The Natural History of Selborne* went through over a hundred reprints before 1955, strongly influencing the early environmental movement (Worster 1991: 7). In Worster's words, he 'grasped the complex unity in diversity that made of the Selborne environs an ecological whole'. Here we find White describing the importance of the humble but vital earthworm in this whole.

Ralph Waldo Emerson (1803–82), the American poet, equally interpreted the complexity and balance of nature as evidence of a creator God. In understanding the web of life that connects species, Emerson reached mystical rather than scientific conclusions. Although statements of his belief in 'an occult relation between man and the vegetable' may sound bizarre or rather floral, his 'transcendentalism', which inspired Thoreau, was based on a solidly ecological outlook. Much of the deep ecology philosophy so popular with later Americans such as Muir and Leopold may also be traced to the influence of Emerson. An arcadian rather than an imperialist, Emerson stressed that 'In the wilderness, I find something more dear and connate than in streets or villages'. In 1833 he travelled to England, where he met Carlyle, Coleridge and Wordsworth. Inspired by the experience, he established a Transcendental Club in 1836 to promote Romantic notions and German philosophy.

Darwin, providing an example of more strictly scientific ecology in the next extract, stated explicitly in his writing the need to understand the 'complex relations of all animals and plants'. Significantly Darwin's theory of evolution was inspired in part by Malthus's interest in population and competition. Both Malthus and Darwin have been used by more conservative commentators to draw conclusions around a postulated 'Social Darwinism' that feared overpopulation and believed none the less in the 'survival of the fittest'. The anarchist Kropotkin's *Mutual Aid* was an attempt to use the examples of ecology to argue that co-operation is just as important as aggressive red-in-tooth-and-claw competition.

Linnaeus

Simple, beautiful, and instructive

Man, when he enters the world, is naturally led to enquire who he is; whence he comes; whither he is going; for what purpose he is created; and by whose benevolence he is preserved. He finds himself descended from the remotest creation; journeying to a life of perfection and happiness; and led by his endowments to a contemplation of the works of nature.

Like other animals who enjoy life, sensation, and perception; who seek for food, amusements, and rest, and who prepare habitations convenient for their kind, he is curious and inquisitive: but, above all other animals, he is noble in his nature, in as much as, by the powers of his mind, he is able to reason justly upon whatever discovers itself to his senses; and to look, with reverence and wonder, upon the works of Him who created all things.

That existence is surely contemptible, which regards only the gratification of instinctive wants, and the preservation of a body made to perish. It is therefore the business of a thinking being, to look forward to the purposes of all things; and to remember that the end of creation is, that God may be glorified in all his works.

Hence it is of importance that we should study the works of nature, than which, what can be more useful, what more interesting? For, however large a portion of them lies open to our present view; a still greater part is yet unknown and undiscovered.

All things are not within the immediate reach of human capacity. Many have been made known to us, of which those who went before us were ignorant; many we have heard of, but know not what they are; and many must remain for the diligence of future ages.

It is the exclusive property of man, to contemplate and to reason on the great book of nature. She gradually unfolds herself to him, who with patience and perseverance, will search into her misteries; and when the memory of the present and of past generations shall be entirely obliterated, he shall enjoy the high privilege of living in the minds of his successors, as he has been advanced in the dignity of his nature, by the labours of those who went before him. . . .

The EARTH is a planetary sphere, turning round its own axis, once in 24 hours, and round the sun once a year; surrounded by an *atmosphere* of elements, and covered by a stupendous crust of *natural bodies*, which are the objects of our studies. It is terraqueous; having the depressed parts covered with waters; the elevated parts gradually dilated into dry and habitable continents. The *land* is moistened by *vapours*, which rising from the waters, are collected into *clouds:* these are deposited upon the tops of mountains; form small *streams*, which unite into *rivulets*, and reunite into those ever-flowing *rivers*, which pervading the thirsty earth, and affording moisture to the productions growing for the support of her living inhabitants, are at last returned into their parent *sea.*

The study of natural history, simple, beautiful, and instructive, consists in the collection, arrangement, and exhibition of the various productions of the earth.

These are divided into the three grand kingdoms of nature, whose boundaries meet together in the Zoophytes.

C. von Linné, *A General System of Nature,* London: Lackington (1806), 1–2.

Gilbert White

Letter XXXV

Dear Sir,   *Selborne*, 20 May 1777

Lands that are subject to frequent inundations are always poor; and probably the reason may be because the worms are drowned. The most insignificant insects and reptiles are of much more consequence, and have much more influence in the œconomy of nature, than the incurious are aware of; and are mighty in their effect, from their minuteness, which renders them less an object of attention; and from their numbers and fecundity. Earth-worms, though in appearance a small and despicable link in the chain of nature, yet, if lost, would make a lamentable chasm. For, to say nothing of half the birds, and some quadrupeds which are almost entirely supported by them, worms seem to be the great promoters of vegetation, which would proceed but lamely without them, by boring, perforating, and loosening the soil, and rendering it pervious to rains and the fibres of plants, by drawing straws and stalks of leaves and twigs into it; and, most of all, by throwing up such infinite numbers of lumps of earth called worm-casts, which, being their excrement, is a fine manure for grain and grass. Worms probably provide new soil for hills and slopes where the rain washes the earth away; and they affect slopes, probably to avoid being flooded. Gardeners and farmers express their detestation of worms; the former because they render their walks unsightly and make them much work: and the latter because, as they think, worms eat their green corn. But these men would find that the earth without worms would soon become cold, hard-bound, and void of fermentation; and consequently sterile: and besides, in favour of worms, it should be hinted that green corn, plants, and flowers, are not so much injured by them as by many species of *coleoptera* (scarabs), and *tipulæ* (long-legs), in their larva, or grub-state; and by unnoticed myriads of small shell-less snails, called slugs, which silently and imperceptibly make amazing havoc in the field and garden.

These hints we think proper to throw out in order to set the inquisitive and discerning to work.

A good monography of worms would afford much entertainment and information at the same time, and would open a large and new field in natural history. Worms work most in the spring; but by no means lie torpid in the dead months; are out every mild night in the winter, as any person may be convinced that will take the pains to examine his grass-plots with a candle, are hermaphrodites, and much addicted to venery, and consequently very prolific.

Gilbert White, *The Natural History of Selborne* (1789), Bristol: Arrowsmith (1924), 181.

Ralph Waldo Emerson

Nature

Our age is retrospective. It builds the sepulchres of the fathers. It writes biographies, histories, and criticism. The foregoing generations beheld God and nature face to face; we, through their eyes. Why should not we also enjoy an original relation to the universe? Why should not we have a poetry and philosophy of insight and not of tradition, and a religion by revelation to us, and not the history of theirs? Embosomed for a season in nature, whose floods of life stream around and through us, and invite us by the powers they supply, to action proportioned to nature, why should we grope among the dry bones of the past, or put the living generation into masquerade out of its faded wardrobe? The sun shines to-day also. There is more wool and flax in the fields. There are new lands, new men, new thoughts. Let us demand our own works and laws and worship.

Undoubtedly we have no questions to ask which are unanswerable. We must trust the perfection of the creation so far as to believe that whatever curiosity the order of things has awakened in our minds, the order of things can satisfy. Every man's condition is a solution in hieroglyphic to those inquiries he would put. He acts it as life, before he apprehends it as truth. In like manner, nature is already, in its forms and tendencies, describing its own design. Let us interrogate the great apparition that shines so peacefully around us. Let us inquire, to what end is nature?

All science has one aim, namely, to find a theory of nature. We have theories of races and of functions, but scarcely yet a remote approach to an idea of creation. We are now so far from the road to truth, that religious teachers dispute and hate each other, and speculative men are esteemed unsound and frivolous. But to a sound judgment, the most abstract truth is the most practical. Whenever a true theory appears, it will be its own evidence. Its test is, that it will explain all phenomena. Now many are thought not only unexplained but inexplicable; as language, sleep, madness, dreams, beasts, sex.

Philosophically considered, the universe is composed of Nature and the Soul. Strictly speaking, therefore, all that is separate from us, all which Philosophy distinguishes as the NOT ME, that is, both nature and art, all other men and my own body, must be ranked under this name, NATURE. In enumerating the values of nature and casting up their sum, I shall use the word in both senses; – in its common and in its philosophical import. In inquiries so general as our present one, the inaccuracy is not material; no confusion of thought will occur. *Nature*, in the common sense, refers to essences unchanged by man; space, the air, the river, the leaf. *Art* is applied to the mixture of his will with the same things, as in a house, a canal, a statue, a picture. But his operations taken together are so insignificant, a little chipping, baking, patching, and washing, that in an impression so grand as that of the world on the human mind, they do not vary the result. . . .

To go into solitude, a man needs to retire as much from his chamber as from society. I am not solitary whilst I read and write, though nobody is with me. But if a man would be alone, let him look at the stars. The rays that come from those heavenly worlds will separate between him and what he touches. One might think the atmosphere was made transparent with this design, to give man, in the heavenly bodies, the perpetual presence of the sublime. Seen in the streets of cities, how great they are! If the stars should appear one night in a thousand years, how would men believe and adore; and preserve for many generations the remembrance of the city of God which had been shown! But every night come out these envoys of beauty, and light the universe with their admonishing smile.

The stars awaken a certain reverence, because though always present, they are inaccessible; but all natural objects make a kindred impression, when the mind is open to their influence. Nature never wears a mean appearance. Neither does the wisest man extort her secret, and lose his curiosity by finding out all her perfection. Nature never became a toy to a wise spirit. The flowers, the animals, the mountains, reflected the wisdom of his best hour, as much as they had delighted the simplicity of his childhood.

When we speak of nature in this manner, we have a distinct but most poetical sense in the mind. We mean the integrity of impression made by manifold natural objects. It is this which distinguishes the stick of timber of the wood-cutter, from the tree of the poet. The charming landscape which I saw this morning is indubitably made up of some twenty or thirty farms. Miller owns this field, Locke that, and Manning the woodland beyond. But none of them owns the landscape. There is a property in the horizon which no man has but he whose eye can integrate all the parts, that is, the poet. This is the best part of these men's farms, yet to this their warranty-deeds give no title.

To speak truly, few adult persons can see nature. Most persons do not see the sun. At least they have a very superficial seeing. The sun illuminates only the eye of the man, but shines into the eye and the heart of the child. The lover of nature is he whose inward and outward senses are still truly adjusted to each other; who has retained the spirit of infancy even into the era of manhood. His intercourse with heaven and earth becomes part of his daily food. In the presence of nature a wild delight runs through the man, in spite of real sorrows. Nature says, – he is my creature, and maugre all his impertinent griefs, he shall be glad with me. Not the sun or the summer alone, but every hour and season yields its tribute of delight; for every hour and change corresponds to and authorizes a different state of the mind, from breathless noon to grimmest midnight. Nature is a setting that fits equally well a comic or a mourning piece. In good health, the air is a cordial of incredible virtue. Crossing a bare common, in snow puddles, at twilight, under a clouded sky, without having in my thoughts any occurrence of special good fortune, I have enjoyed a perfect exhilaration. I am glad to the brink of fear. In the woods, too, a man casts off his years, as the snake his slough, and at what period soever of life, is always a child. In the woods is perpetual youth. Within these plantations of God, a decorum and sanctity reign, a perennial festival is dressed, and the guest sees not how he should tire of them in a thousand years. In the woods, we return to reason and faith. There I feel that nothing can befall me in life, – no disgrace, no calamity (leaving me my eyes), which nature cannot repair. Standing on the bare ground, – my head bathed by the blithe air, and uplifted into infinite space, – all mean egotism vanishes. I become a transparent eye-ball; I am nothing; I see all; the currents of the Universal Being circulate through me; I am part or parcel of God. The name of the nearest friend sounds then foreign and accidental: to be brothers, to be acquaintances, – master or servant, is then a trifle and a disturbance. I am the lover of uncontained and immortal beauty. In the wilderness, I find something more dear and connate than in streets or villages. In the tranquil landscape, and especially in the distant line of the horizon, man beholds somewhat as beautiful as his own nature.

The greatest delight which the fields and woods minister is the suggestion of an occult relation between man and the vegetable. I am not alone and unacknowledged. They nod to me, and I to them. The waving of the boughs in the storm is new to me and old. It takes me by surprise, and yet is not unknown. Its effect is like that of a higher thought or a better emotion coming over me, when I deemed I was thinking justly or doing right.

Yet it is certain that the power to produce this delight does not reside in nature, but in man, or in a harmony of both. It is necessary to use these pleasures with great temperance. For nature is not always tricked in holiday attire, but the same scene which yesterday breathed perfume and glittered as for the frolic of the nymphs, is overspread with melancholy to-day. Nature always wears the colors of the spirit. To a man laboring under calamity, the heat of his own fire hath sadness in it. Then there is a kind of contempt of the landscape felt by him who has just lost by death a dear friend. The sky is less grand as it shuts down over less worth in the population.

Nature: Addresses and Lectures (1836), Boston; Mass.: Houghton Mifflin (1883), 9–17.

Charles Darwin

The complex relations of all animals and plants

When a species, owing to highly favourable circumstances, increases inordinately in numbers in a small tract, epidemics – at least, this seems generally to occur with our game animals – often ensue; and here we have a limiting check independent of the struggle for life. But even some of these so-called epidemics appear to be due to parasitic worms, which have from some cause, possibly in part through facility of diffusion amongst the crowded animals, been disproportionally favoured: and here comes in a sort of struggle between the parasite and its prey.

On the other hand, in many cases, a large stock of individuals of the same species, relatively to the numbers of its enemies, is absolutely necessary for its preservation. Thus we can easily raise plenty of corn and rape-seed, &c., in our fields, because the seeds are in great excess compared with the number of birds which feed on them; nor can the birds, though having a superabundance of food at this one season, increase in number proportionally to the supply of seed, as their numbers are checked during winter; but any one who has tried, knows how troublesome it is to get seed from a few wheat or other such plants in a garden: I have in this case lost every single seed. This view of the necessity of a large stock of the same species for its preservation, explains, I believe, some singular facts in nature such as that of very rare plants being sometimes extremely abundant, in the few spots where they do exist; and that of some social plants being social, that is abounding in individuals, even on the extreme verge of their range. For in such cases, we may believe, that a plant could exist only where the conditions of its life were so favourable that many could exist together, and thus save the species from utter destruction. I should add that the good effects of intercrossing, and the ill effects of close interbreeding, no doubt come into play in many of these cases; but I will not here enlarge on this subject.

Complex Relations of all Animals and Plants to each other in the Struggle for Existence.

Many cases are on record showing how complex and unexpected are the checks and relations between organic beings, which have to struggle together in the same country. I will give only a single instance, which, though a simple one, interested me. In Stafford-shire, on the estate of a relation, where I had ample means of investigation, there was a large and extremely barren heath, which had never been touched by the hand of man; but several acres of exactly the same nature had been enclosed twenty-five years previously and planted with Scotch fir. The change in the native vegetation of the planted part of the heath was most remarkable, more than is generally seen in passing from one quite different soil to another: not only the proportional numbers of the heath-plants were wholly changed, but twelve species of plants (not counting grasses and carices) flourished in the plantations, which could not be found on the heath. The effect on the insects must have been still greater, for six insectivorous birds were very common in the plantations, which were not to be seen on the heath; and the heath was frequented by two or three distinct insectivorous birds. Here we see how potent has been the effect of the intro-duction of a single tree, nothing whatever else having been done, with the exception of the land having been enclosed, so that cattle could not enter. But how important an element enclosure is, I plainly saw near Farnham, in Surrey. Here there are extensive heaths, with a few clumps of old Scotch firs on the distant hill-tops: within the last ten years large spaces have been enclosed, and self sown firs are now springing up in multitudes, so close together that all cannot live. When I ascertained that these young trees had not been sown or planted, I was so much surprised at their numbers that I went to several points of view, whence I could examine hundreds of acres of the unenclosed heath, and literally I could not see a single Scotch fir, except the old planted clumps. But on looking closely between the stems of the heath, I found a multitude of seedlings and little trees which had been perpetually browsed down by the cattle. In one square yard, at a point some hundred yards distant from one of the old clumps, I counted thirty-two little trees; and one of them, with twenty-six rings of growth, had, during many years, tried to raise its head above the stems of the heath, and had failed. No wonder that, as soon as the land was enclosed, it became thickly clothed with vigorously growing young firs. Yet the heath was so extremely barren and so extensive that no one would ever have imagined that cattle would have so closely and effectually searched it for food.

Here we see that cattle absolutely determine the existence of the Scotch fir; but in several parts of the world insects determine the existence of cattle. Perhaps Paraguay offers the most curious instance of this; for here neither cattle nor horses nor dogs have ever run wild, though they swarm southward and northward in a feral state; and Azara and Rengger have shown that this is caused by the greater number in Paraguay of a certain fly, which lays its eggs in the navels of these animals when first born. The increase of these flies, numerous as they are, must be habitually checked by some means, probably by other parasitic insects. Hence, if certain insectivorous birds were to decrease in Paraguay, the parasitic insects would probably increase; and this would lessen the number of the navel-frequenting flies – then cattle and horses would become feral, and this would certainly greatly alter (as indeed I have observed in parts of South America) the vegeta-tion: this again would largely affect the insects; and this, as we have just seen in Stafford-shire, the insectivorous birds, and so onwards in ever-increasing circles of complexity. Not that under nature the relations will ever be as simple as this. Battle within battle must be continually recurring with varying success; and yet in the long run the forces are so nicely balanced, that the face of nature remains for long periods of time uniform, though assuredly the merest trifle would give the victory to one organic being over another. Nevertheless, so profound is our ignorance, and so high our presumption, that we marvel when we hear of the extinction of an organic being; and as we do not see the cause, we invoke cataclysms to desolate the world, or invent laws on the duration of the forms of life!

I am tempted to give one more instance showing how plants and animals, remote in the scale of nature, are bound together by a web of complex relations. I shall hereafter have occasion to show that the exotic Lobelia fulgens is never visited in my garden by insects, and consequently, from its peculiar structure, never sets a seed. Nearly all our orchid-aceous plants absolutely require the visits of insects to remove their pollen-masses and thus to fertilise them. I find from experiments that humble-bees are almost indispensable to the fertilisation of the heartsease (Viola tricolor), for other bees do not visit this flower. I have also found that the visits of bees are necessary for the fertilisation of some kinds of clover; for instance, 20 heads of Dutch clover (Trifolium repens) yielded 2,290 seeds, but 20 other heads protected from bees produced not one. Again, 100 heads of red clover (T. pratense) produced 2,700 seeds, but the same number of protected heads produced not a single seed. Humble-bees alone visit red clover, as other bees cannot reach the nectar. It has been suggested that moths may fertilise the clovers; but I doubt whether they could do so in the case of the red clover, from their weight not being sufficient to depress the wing petals. Hence we may infer as highly probable that, if the whole genus of humble-bees became extinct or very rare in England, the heartsease and red clover would become very rare, or wholly disappear. The number of humble-bees in any district depends in a great measure upon the number of field-mice, which destroy their combs and nests; and Col. Newman, who has long attended to the habits of humble-bees, believes that 'more than two-thirds of them are thus destroyed all over England'. Now the number of mice is largely dependent, as every one knows, on the number of cats....

On the Origin of Species, London: Murray (1859), 51–3

Peter Kropotkin

Mutual aid

Two aspects of animal life impressed me most during the journeys which I made in my youth in Eastern Siberia and Northern Manchuria. One of them was the extreme severity of the struggle for existence which most species of animals have to carry on against an inclement Nature; the enormous destruction of life which periodically results from natural agencies; and the consequent paucity of life over the vast territory which fell under my observation. And the other was, that even in those few spots where animal life teemed in abundance, I failed to find – although I was eagerly looking for it – that bitter struggle for the means of existence, *among animals belonging to the same species,* which was considered by most Darwinists (though not always by Darwin himself) as the dominant characteristic of struggle for life, and the main factor of evolution.

The terrible snow-storms which sweep over the northern portion of Eurasia in the later part of the winter, and the glazed frost that often follows them; the frosts and the snow-storms which return every year in the second half of May, when the trees are already in full blossom and insect life swarms everywhere; the early frosts and, occasionally, the heavy snowfalls in July and August, which suddenly destroy myriads of insects, as well as the second broods of the birds in the prairies; the torrential rains, due to the monsoons, which fall in more temperate regions in August and September – resulting in inundations on a scale which is only known in America and in Eastern Asia, and swamping, on the plateaus, areas as wide as European States; and finally, the heavy snowfalls, early in October, which eventually render a territory as large as France and Germany, absolutely impracticable for ruminants, and destroy them by the thousand – these were the conditions under which I saw animal life struggling in Northern Asia. They made me realize at an early date the overwhelming importance in nature of what Darwin described as 'the natural checks to over-multiplication', in comparison to the struggle between individuals of the same species for the means of subsistence, which may go on here and there, to some limited extent, but never attains the importance of the former. Paucity of life, under-population – not over-population – being the distinctive feature of that immense part of the globe which we name Northern Asia, I conceived since then serious doubts – which subsequent study has only confirmed – as to the reality of that fearful competition for food and life within each species, which was an article of faith with most Darwinists, and, consequently, as to the dominant part which this sort of competition was supposed to play in the evolution of new species.

On the other hand, wherever I saw animal life in abundance, as, for instance, on the lakes where scores of species and millions of individuals came together to rear their progeny; in the colonies of rodents; in the migrations of birds which took place at that time on a truly American scale along the Usuri; and especially in a migration of fallow-deer which I witnessed on the Amur, and during which scores of thousands of these intelligent animals came together from an immense territory, flying before the coming deep snow, in order to cross the Amur where it is narrowest – in all these scenes of animal life which passed before my eyes, I saw mutual aid and mutual support carried on to an extent which made me suspect in it a feature of the greatest importance for the maintenance of life, the preservation of each species, and its further evolution.

And finally, I saw among the semi-wild cattle and horses in Transbaikalia, among the wild ruminants everywhere, the squirrels, and so on, that when animals have to struggle against scarcity of food, in consequence of one of the above-mentioned causes, the whole of that portion of the species which is affected by the calamity, comes out of the ordeal so much impoverished in vigour and health, that *no progressive evolution of the species can be based upon such periods of keen competition.*

Consequently, when my attention was drawn, later on, to the relations between

Darwinism and sociology, I could agree with none of the works and pamphlets that had been written upon this important subject. They all endeavoured to prove that man, owing to his higher intelligence and knowledge, *may* mitigate the harshness of the struggle for life between men; but they all recognized at the same time that the struggle for the means of existence, of every animal against all its congeners, and of every man against all other men, was 'a law of Nature'. This view, however, I could not accept, because I was persuaded that to admit a pitiless inner war for life within each species, and to see in that war a condition of progress, was to admit something which not only had not yet been proved, but also lacked confirmation from direct observation.

Mutual Aid (1914), Boston; Mass.: Horizon (1955), 17–19.

CHAPTER 9

Against growth

T. R. Malthus, The principle of population

J. de Sismondi, Political economy and principles of government

J. S. Mill, The stationary state

J. Ruskin, Unto this last

A. Huxley, Progress!

Introduction

Franklin shook his head, 'People won't stand for it.'

'They will. Within the last twenty-five years the gross national product has risen by fifty per cent, but so have the average hours worked. Ultimately we'll all be working and spending twenty-four hours a day, seven days a week. Think what a slump would mean – millions of lay-offs, people with time on their hands and nothing to spend it on. Real leisure, not just time spent buying things.'

(Ballard 1969: 68)

The most striking and perhaps unpalatable part of the Green message has been a rejection of both economic growth and growth in human numbers. Greens are opposed to simple notions of 'progress' and economic expansion. While other politicians vie with one another to maximize the standard of living, Green aspirants to office cycle in the opposite direction. The contrast between simple environmental concern and a critique that demands a static or diminished economic basis to maintain global ecology is fundamental. The growth test may be used as a swift and sure way of separating Green ideology from environmentalism.

Curiously, perhaps, calls for 'zero growth' have been raised by economists and others for over 200 years. The substance of the modern Green critique is clearly indicated in our historical sources. The Right Reverend Thomas Robert Malthus (1766–1834), perhaps, needs no introduction. His claim that populations could not increase without eventually outstripping their resource base and creating starvation is infamous. Critics, including Cobbett, Godwin, Marx and other radicals, accused him of using his theory to suggest that the poverty of his day, far from being caused by social injustice, was an inevitable product of natural process. Benton (1989) postulates that Malthus, the son of a radical, familiar with the utopian writings of Godwin, was an 'epistemic' conservative, i.e. his conservatism came not from enthusiasm for the inequality and poverty of his age but from the belief that such suffering was inevitable. He was conservative of necessity rather than choice, if we are to believe Benton. Malthus would rather have agitated with the likes of Shelley and Cobbett. The political ecologists of the 1970s and elements of the modern Green movement can clearly be labelled neo-Malthusian. Yet the description 'neo-Shelleyan' might be more accurate, for did not the poet claim that arguments against unlimited expansion demand not the preservation of inequality but its abolition. In a society where the economy no longer grows, the case for sharing wealth evenly increases

116

proportionately. Shelley, as we shall see (Chapter Ten), together with his father-in-law, William Godwin, argued that vegetarianism would allow a more sustainable economy and help curb the power of those with wealth.

Jean de Sismondi (1773–1842), the Swiss economist who inspired Marx and prefigured Keynes with his concept of cyclical economic depression caused by overproduction, is another source of early anti-growth sentiment. He rejected unrestrained industrial expansion as an essentially anti-human trend. Observing at first hand the British industrial revolution, he noted, 'In this astonishing country ... I have seen production increasing whilst enjoyments were diminishing.' Condemned as a backward-looking Romantic by Marx, he none the less reminds us that 'the increase of wealth is not the end in political economy, but its instrument in procuring the happiness of all'. However we analyse the contribution and political motivation of Malthus, in Sismondi's case it is clear that his critique of growth was motivated by humanism and a concern for environmental quality.

Whilst Malthus was a conservative and other of our writers tend towards the left, John Stuart Mill (1806–73) is often seen as the main inspiration of modern liberalism. He rejects an economy based upon 'the trampling, crushing, elbowing, and treading ...' of each individual. He reminds us that 'a stationary condition of capital and population implies no stationary state of human improvement'. He even touches on deep Green themes of wilderness preservation and hints at animal rights, condemning a world where 'all quadrupeds or birds which are not domesticated for man's use [are] exterminated as his rivals for food'.

John Ruskin (1819–1900), the art critic and inspirer of William Morris, wrote influentially on political economy. He argued that not just production but the provision of good and satisfying work was important. In this passage from his essay *Unto this Last* he describes how the world 'cannot become a factory or a mine'. He dwells on the conflict between economic expansion and the need to preserve nature. Finally, in 1928, we find Huxley's (1894–1963) Lord Edward lambasting a would-be British fascist leader, in the novel *Point Counter Point*. 'Progress' acts both for his lordship and for modern Greens as an excuse for waste, poor economic management and the drive to wreck the planet's life systems.

Thomas Robert Malthus

The principle of population

I think I may fairly make two postulata.

First, That food is necessary to the existence of man.

Secondly, That the passion between the sexes is necessary and will remain nearly in its present state.

These two laws, ever since we have had any knowledge of mankind, appear to have been fixed laws of our nature, and, as we have not hitherto seen any alteration in them, we have no right to conclude that they will ever cease to be what they now are, without an immediate act of power in that Being who first arranged the system of the universe, and for the advantage of his creatures, still executes, according to fixed laws, all its various operations.

I do not know that any writer has supposed that on this earth man will ultimately be able to live without food. But Mr Godwin has conjectured that the passion between the sexes may in time be extinguished. As, however, he calls this part of his work a deviation into the land of conjecture, I will not dwell longer upon it at present than to say that the best arguments for the perfectibility of man are drawn from a contemplation of the great progress that he had already made from the savage state and the difficulty of saying where he is to stop. But towards the extinction of the passion between the sexes, no progress whatever has hitherto been made. It appears to exist in as much force at present as it did two thousand or four thousand years ago. There are individual exceptions now as there always have been. But, as these exceptions do not appear to increase in number, it would surely be a very unphilosophical mode of arguing to infer, merely from the existence of an exception, that the exception would, in time, become the rule, and the rule the exception.

Assuming then my postulata as granted, I say, that the power of population is indefinitely greater than the power in the earth to produce subsistence for man.

Population, when unchecked, increases in a geometrical ratio. Subsistence increases only in an arithmetical ratio. A slight acquaintance with numbers will shew the immensity of the first power in comparison of the second.

By that law of our nature which makes food necessary to the life of man, the effects of these two unequal powers must be kept equal.

This implies a strong and constantly operating check on population from the difficulty of subsistence. This difficulty must fall somewhere and must necessarily be severely felt by a large portion of mankind. . . .

. . . The germs of existence contained in this spot of earth, with ample food, and ample room to expand in, would fill millions of worlds in the course of a few thousand years. Necessity, that imperious all pervading law of nature, restrains them within the prescribed bounds. The race of plants and the race of animals shrink under this great restrictive law. And the race of man cannot, by any efforts of reason, escape from it. Among plants and animals its effects are waste of seed, sickness, and premature death. Among mankind, misery and vice. The former, misery, is an absolutely necessary consequence of it. Vice is a highly probable consequence, and we therefore see it abundantly prevail, but it ought not, perhaps, to be called an absolutely necessary consequence. The ordeal of virtue is to resist all temptation to evil.

This natural inequality of the two powers of population and of production in the earth, and that great law of our nature which must constantly keep their effects equal, form the great difficulty that to me appears insurmountable in the way to the perfectibility of society. All other arguments are of slight and subordinate consideration in comparison of this. I see no way by which man can escape from the weight of this law which pervades all animated nature. No fancied equality, no agrarian regulations in their utmost extent, could remove the pressure of it even for a single century. And it appears, therefore, to be decisive against the possible existence of a society, all the members of which should live in ease, happiness, and comparative leisure; and feel no anxiety about providing the means of subsistence for themselves and families.

Consequently, if the premises are just, the argument is conclusive against the perfectibility of the mass of mankind.

I have thus sketched the general outline of the argument, but I will examine it more particularly, and I think it will be found that experience, the true source and foundation of all knowledge, invariably confirms its truth.

On the Principle of Population (1798), Harmondsworth: Penguin (1970) 70–2.

Jean de Sismondi

Political economy and principles of government

The study of England has confirmed me in my 'New Principles'. In this astonishing country, which seems to be submitted to a great experiment for the instruction of the rest of the world, I have seen production increasing whilst enjoyments were diminishing. The mass of the nation here, no less than philosophers, seems to forget that the increase of wealth is not the end in political economy, but its instrument in procuring the happiness of all. I sought for this happiness in every class, and I could nowhere find it. . . .

One has great reason to be astonished that a system which tends to destroy small properties in mechanical arts, as well as in agriculture, and to substitute for them, indigence on one side and opulence on the other – a system which creates for some, power unbounded, and for others absolute dependence – a system whose tendency is in opposition to the governing idea and passion of the age, equality – should, exactly in such an age, have been received with so much favour. Nevertheless, it is a fact; and industrialism has been proclaimed as being the tendency, as well as the glory of our age, by persons who would equally have paid homage to the feudalism of the twelfth century. Still more, the Saint-Simonians have almost made a religion of it. The prodigies of the mind of man subduing the elements have made them forget that the character of true prodigies, of those that come from God, is to be useful to man.

New Principles of Political Economy (1819), London: Chapman (1847), 115, 205–6.

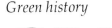

John Stuart Mill

The stationary state

§ 2. I cannot, therefore, regard the stationary state of capital and wealth with the unaffected aversion so generally manifested towards it by political economists of the old school. I am inclined to believe that it would be, on the whole, a very considerable improvement on our present condition. I confess I am not charmed with the ideal of life held out by those who think that the normal state of human beings is that of struggling to get on; that the trampling, crushing, elbowing, and treading on each other's heels, which form the existing type of social life, are the most desirable lot of human kind, or anything but the disagreeable symptoms of one of the phases of industrial progress. It may be a necessary stage in the progress of civilization, and those European nations which have hitherto been so fortunate as to be preserved from it, may have it yet to undergo. It is an incident of growth, not a mark of decline, for it is not necessarily destructive of the higher aspirations and the heroic virtues; as America, in her great civil war, has proved to the world, both by her conduct as a people and by numerous splendid individual examples, and as England, it is to be hoped, would also prove, on an equally trying and exciting occasion. But it is not a kind of social perfection which philanthropists to come will feel any very eager desire to assist in realizing. Most fitting, indeed, is it, that while riches are power, and to grow as rich as possible the universal object of ambition, the path to its attainment should be open to all, without favour or partiality. But the best state for human nature is that in which, while no one is poor, no one desires to be richer, nor has any reason to fear being thrust back, by the efforts of others to push themselves forward.

That the energies of mankind should be kept in employment by the struggle for riches, as they were formerly by the struggle of war, until the better minds succeed in educating the others into better things, is undoubtedly more desirable than that they should rust and stagnate. While minds are coarse they require coarse stimuli, and let them have them. In the meantime, those who do not accept the present very early stage of human improve-ment as its ultimate type, may be excused for being comparatively indifferent to the kind of economical progress which excites the congratulations of ordinary politicians; the mere increase of production and accumulation. For the safety of national independence it is essential that a country should not fall much behind its neighbours in these things. But in themselves they are of little importance, so long as either the increase of population or anything else prevents the mass of the people from reaping any part of the benefit of them. I know now why it should be matter of congratulation that persons who are already richer than any one needs to be, should have doubled their means of consuming things which give little or no pleasure except as representative of wealth; or that numbers of individuals should pass over, every year, from the middle classes into a richer class, or from the class of the occupied rich to that of the unoccupied. It is only in the backward countries of the world that increased production is still an important object: in those most advanced, what is economically needed is a better distribution, of which one indispensable means is a stricter restraint on population. Levelling institutions, either of a just or of an unjust kind, cannot alone accomplish it; they may lower the heights of society, but they cannot, of themselves, permanently raise the depths.

On the other hand, we may suppose this better distribution of property attained, by the joint effect of the prudence and frugality of individuals, and of a system of legislation favouring equality of fortunes, so far as is consistent with the just claim of the individual to the fruits, whether great or small, of his or her own industry. We may suppose, for instance (according to the suggestion thrown out in a former chapter), a limitation of the sum which any one person may acquire by gift or inheritance, to the amount sufficient to constitute a moderate independence. Under this twofold influence, society would exhibit these leading features: a well paid and affluent body of labourers; no enormous fortunes, except what were earned and accumulated during a single lifetime; but a much larger body of persons than at present, not only exempt from the coarser toils, but with sufficient leisure, both physical and mental, from mechanical details, to cultivate freely the graces of life, and afford examples of them to the classes less favourably circumstanced for their growth. This condition of society, so greatly preferable to the present, is not only perfectly compatible with the stationary state, but, it would seem, more naturally allied with that state than with any other.

There is room in the world, no doubt, and even in old countries, for a great increase of population, supposing the arts of life to go on improving, and capital to increase. But even if innocuous, I confess I see very little reason for desiring it. The density of population necessary to enable mankind to obtain, in the greatest degree, all the advantages both of co-operation and of social intercourse, has, in all the most populous countries, been attained. A population may be too crowded, though all be amply supplied with food and raiment. It is not good for man to be kept perforce at all times in the presence of his species. A world from which solitude is extirpated, is a very poor ideal. Solitude, in the sense of being often alone, is essential to any depth of meditation or of character; and solitude in the presence of natural beauty and grandeur, is the cradle of thoughts and aspirations which are not only good for the individual, but which society could ill do without. Nor is there much satisfaction in contemplating the world with nothing left to the spontaneous activity of nature; with every rood of land brought into cultivation, which is capable of growing food for human beings; every flowery waste or natural pasture ploughed up, all quadrupeds or birds which are not domesticated for man's use exterminated as his rivals for food, every hedgerow or superfluous tree rooted out, and scarcely a place left where a wild shrub or flower could grow without being eradicated as a weed in the name of improved agriculture. If the earth must lose that great portion of its pleasantness which it owes to things that the unlimited increase of wealth and population would extirpate from it, for the mere purpose of enabling it to support a larger, but not a better or a happier population, I sincerely hope, for the sake of posterity, that they will be content to be stationary, long before necessity compels them to it.

It is scarcely necessary to remark that a stationary condition of capital and population implies no stationary state of human improvement. There would be as much scope as ever for all kinds of mental culture, and moral and social progress; as much room for improving the Art of Living, and much more likelihood of its being improved, when minds ceased to be engrossed by the art of getting on.

Principles of Political Economy (1848), London: Longman (1871), 2: 328–32.

John Ruskin

Unto this last

Leaving these questions to be discussed, or waived, at their pleasure, by Mr. Ricardo's followers, I proceed to state the main facts bearing on that probable future of the labouring classes which has been partially glanced at by Mr. Mill. That chapter and the preceding one differ from the common writing of political economists in admitting some value in the aspect of nature, and expressing regret at the probability of the destruction of natural scenery. But we may spare our anxieties on this head. Men can neither drink steam, nor eat stone. The maximum of population on a given space of land implies also the relative maximum of edible vegetable, whether for men or cattle; it implies a maximum of pure air; and of pure water. Therefore: a maximum of wood, to transmute the air, and of sloping ground, protected by herbage from the extreme heat of the sun, to feed the streams. All England may, if it so chooses, become one manufacturing town; and Englishmen, sacrificing themselves to the good of general humanity, may live diminished lives in the midst of noise, of darkness, and of deadly exhalation. But the world cannot become a factory, nor a mine. No amount of ingenuity will ever make iron digestible by the million, nor substitute hydrogen for wine. Neither the avarice nor the rage of men will ever feed them; and however the apple of Sodom and the grape of Gomorrah may spread their table for a time with dainties of ashes, and nectar of asps, – so long as men live by bread, the far away valleys must laugh as they are covered with the gold of God, and the shouts of His happy multitudes ring round the winepress and the well.

Nor need our more sentimental economists fear the too wide spread of the formalities of a mechanical agriculture. The presence of a wise population implies the search for felicity as well as for food; nor can any population reach its maximum but through that wisdom which 'rejoices' in the habitable parts of the earth. The desert has its appointed place and work; the eternal engine, whose beam is the earth's axle, whose beat is its year, and whose breath is its ocean, will still divide imperiously to their desert kingdoms bound with unfurrowable rock, and swept by unarrested sand, their powers of frost and fire: but the zones and lands between, habitable, will be loveliest in habitation. The desire of the heart is also the light of the eyes. No scene is continually and untiringly loved, but one rich by joyful human labour; smooth in field; fair in garden; full in orchard; trim, sweet, and frequent in homestead; ringing with voices of vivid existence. No air is sweet that is silent; it is only sweet when full of low currents of under sound – triplets of birds, and murmur and chirp of insects, and deep-toned words of men, and wayward trebles of childhood. As the art of life is learned, it will be found at last that all lovely things are also necessary: – the wild flower by the wayside, as well as the tended corn; and the wild birds and creatures of the forest, as well as the tended cattle; because man doth not live by bread only, but also by the desert manna; by every wondrous word and unknowable work of God. Happy, in that he knew them not, nor did his fathers know; and that round about him reaches yet into the infinite, the amazement of his existence.

Unto this Last (1861) Orpington: George Allen (1887), 165–8.

Against growth

Aldous Huxley

Progress!

Lord Edward started at the word. It touched a trigger, it released a flood of energy. 'Progress!' he echoed, and the tone of misery and embarrassment was exchanged for one of confidence. 'Progress! You politicians are always talking about it. As though it were going to last. Indefinitely. More motors, more babies, more food, more advertising, more money, more everything, for ever. You ought to take a few lessons in my subject. Physical biology. Progress, indeed! What do you propose to do about phosphorus, for example?' His question was a personal accusation.

'But all this is entirely beside the point,' said Webley impatiently.

'On the contrary,' retorted Lord Edward, 'it's the only point.' His voice had become loud and severe. He spoke with a much more than ordinary degree of coherence. Phosphorus had made a new man of him; he felt very strongly about phosphorus and, feeling strongly, he was strong. The worried bear had become the worrier. 'With your intensive agriculture,' he went on, 'you're simply draining the soil of phosphorous. More than half of one per cent a year. Going clean out of circulation. And then the way you throw away hundreds of thousands of tons of phosphorus pentoxide in your sewage! Pouring it into the sea. And you call that progress. Your modern sewage systems!' His tone was witheringly scornful. 'You ought to be putting it back where it came from. On the land.' Lord Edward shook an admonitory finger and frowned. 'On the land, I tell you.'

'But all this has nothing to do with me,' protested Webley.

'Then it ought to,' Lord Edward answered sternly. 'That's the trouble with you politicians. You don't even think of the important things. Talking about progress and votes and Bolshevism and every year allowing a million tons of phosphorus pentoxide to run away into the sea. It's idiotic, it's criminal, it's . . . it's fiddling while Rome is burning.' He saw Webley opening his mouth to speak and made haste to anticipate what he imagined was going to be his objection. 'No doubt,' he said, 'you think you can make good the loss with phosphate rocks. But what'll you do when the deposits are exhausted?' He poked Everard in the shirt front. 'What then? Only two hundred years and they'll be finished. You think we're being progressive because we're living on our capital. Phosphates, coal, petroleum, nitre – squander them all. That's your policy. And meanwhile you go round trying to make our flesh creep with talk about revolutions.'

'But damn it all,' said Webley, half angry, half amused, 'your phosphorus can wait. This other danger's imminent. Do you *want* a political and social revolution?'

'Will it reduce the population and check production?' asked Lord Edward.

'Of course.'

'Then certainly I want a revolution.' The Old Man thought in terms of geology and was not afraid of logical conclusions. 'Certainly.' Illidge could hardly contain his laughter.

'Well, if that's your view . . .' began Webley; but Lord Edward interrupted him.

'The only result of your progress' he said, 'will be that in a few generations there'll be a real revolution – a natural, cosmic revolution. You're upsetting the equilibrium. And in the end, nature will restore it. And the process will be very uncomfortable for you. Your decline will be as quick as your rise. Quicker, because you'll be bankrupt, you'll have squandered your capital. It takes a rich man a little time to realize all his resources. But when they've all been realized, it takes him almost no time to starve.'

Webley shrugged his shoulders. 'Dotty old lunatic!' he said to himself, and aloud, 'Parallel straight lines never meet, Lord Edward. So I'll bid you good-night.' He took his leave.

A minute later the Old Man and his assistant were making their way up the triumphal staircase to their world apart.

'What a relief!' said Lord Edward, as he opened the door of his laboratory. Voluptuously, he sniffed the faint smell of the absolute alcohol in which the specimens were pickled. 'These parties! One's thankful to get back to science. Still, the music was really . . .' His admiration was inarticulate.

Point Counter Point (1928), London: Chatto & Windus (1971), 78–80.

CHAPTER 10

Sustainable development

J. Evelyn, Sylva

P. B. Shelley, A vindication of natural diet

F. Fourier, The economics of harmony

G. P. Marsh, Recycling and destruction

G. Pinchot, Conservation

Introduction

> Evelyn is as good as several old druids, and his 'Silva' is a new kind of prayer book, a glorifying of the trees and enjoying them forever, which was the chief end of his life.
>
> (Henry Thoreau, quoted in Worster 1991: 87)

If an assault on growth is one half of the Green economic agenda, sustainability, providing an alternative to constant expansion, completes the whole. Economic systems should be infinitely sustainable, cyclical in nature and able to recycle energy and resource inputs. Rather than being based on quantitative measures of gross national product, their goals should be ecologically centred and qualitative. Above all, preservation of, and interaction with, nature are vital. The reduction of human wants and the abolition of degrading, alienating work are also sought. Social justice and the creation of a sense of community are equally important; the end goal of a sustainable economy may in a sense be the abolition of economics as a category separate from other areas of life. In contrast to conventional development theory, the concept of underdeveloped 'third world' nations in need of 'aid' (both intellectual and financial) from a more advanced 'first world' is rejected. Traditional economies, operating largely outside the market, using barter or gift exchange and producing a diversity of crops and products on a local level, had many virtues. Apparent backwardness and underdevelopment have been caused not by lack of technology but by systems of colonial exploitation (Trainer. 1985).

Our final commentator, Gifford Pinchot (1865–1946), a conservationist rather than a deep Green, writing during the Progressive era of US politics, captures the notion of equitable and sustainable economics shared by all Greens. For him:

> The first principle of conservation is development … the use of natural resources now existing for the benefit of the people who live here now … natural resources must be developed and preserved for the benefit of the many and not merely for the profit of the few.

Such a view also reflects the thinking of our first writer, John Evelyn, whom we saw earlier attacking the problem of air pollution and suggesting the application of preventive legislation to his monarch and mentor, Charles II. Both Pinchot and Evelyn (despite Thoreau's praise) are conservationists who stress sound economic management of the resources of nature necessary to the maintenance of society. Radical Greens would attack the implied idea that nature is a resource for human use. The difference between efficient or sustainable exploitation and a deeper Green notion of developing 'non-violent' economic systems that minimize the human impact on other species should be noted. Like modern-day environmentalists bemoaning the fate of the rain forests, Evelyn maintained that trees preserve the climatic balance necessary for human prosperity; in particular encouraging cloud formation, 'so that if their woods were once destroyed, they might perish for want of rain'. The main motive for planting oaks in Evelyn's age was to provide the navy with timber for new ships rather than to preserve nature's beauty.

More radically Shelley, in his vegetarian manifesto, notes that 'The change which would be produced by simpler habits on political economy would be sufficiently remarkable'. He argues, as do many modern Greens, that a shift to a meatless diet would allow the production of far greater quantities of food per acre than a carnivorous diet. Thus 'The monopolising eater of animal flesh would no longer destroy his constitution by devouring an acre at a meal' whilst causing 'the long-protracted famine of the hard-working peasant's hungry babes'. Shelley's 'natural diet' would benefit both nature and those who lived in poverty. Like Gandhi and Morris he argues that human greed should be reduced so that essential needs can be met. His prescription for a just economy pre-figures much modern radical Green sentiment.

The French utopian socialist François Fourier (1772–1837) illustrates another fundamental principle of sustainable development; goods are made to last as long as possible, and all wasteful forms of production are rejected. 'We shall find among the Harmonians', writes Fourier, 'a policy totally contrary to our ideas of commerce, which promote waste and the changes of fashion, under the pretext of maintaining the workman'. George Perkins Marsh (1801–82) echoes Fourier's belief that 'God does not waste an atom' when he states that reckless waste is devastating the planet. His arguments in praise of recycling illustrate that this key demand of ecological economics also has a history. No anthology of Green or environmental ideas would of course be complete without a mention of Marsh, whose volume *Man and Nature*, produced in 1864, was perhaps the first complete publication to criticize in detail human misuse of the Earth.

Other elements of sustainable economics can be found in Chapter Thirteen, which examines Green conceptions of the city and the countryside. Cobbett, Gandhi and contemporary Greens such as Trainer argue that citizens should be intimately involved in food production and cottage industry. Perhaps the Green economy is, in Cobbett's words, a 'cottage economy'.

John Evelyn

Sylva

In the mean time, this of the soil, (which I think is a more proper term for composts) or mould rather, being of greater importance for the raising, planting, and propagation of trees in general, must at no hand be neglected, and is therefore on all occasions mentioned in almost every chapter of our ensuing discourse; I shall therefore not need to assign it any part, when I have affirm'd in general, that most timber-trees grow and prosper well in any tolerable land which will produce corn or rye, and which is not in excess stony; in which nevertheless there are some trees delight; or altogether clay, which few, or none do naturally affect; and yet the oak is seen to prosper in it, for its toughness preferr'd before any other by many workmen, though of all soils the cow-pasture doth certainly exceed, be it for what purpose soever of planting wood. Rather therefore we should take notice how many great wits and ingenious persons, who have leisure and faculty, are in pain for improvements of their heaths and barren Hills, cold and starving places, which causes them to be neglected and despair'd of; whilst they flatter their hopes and vain expectations with fructifying liquors, chymical menstruums, and such vast conceptions; in the mean time that one may shew them as heathy and hopeless grounds, and barren hills as any in England, that do now bear, or lately have born woods, groves, and copses, which yield the owners more wealth, than the richest and most opulent wheat-lands: and if it be objected that 'tis so long a day before these plantations can afford that gain; the Brabant Nurseries, and divers home-plantations of industrious persons are sufficient to convince the gain-sayer. And when by this husbandry a few acorns shall have peopl'd the neighbouring regions with young stocks and trees; the residue will become groves and copses of infinite delight and satisfaction to the planters. Besides, we daily see what course lands will bear these stocks (suppose them oaks, wall-nuts, chess-nuts, pines, firr, ash, wild-pears, crabs, &c.) and some of them (as for instance the pear and the firr or pine) strike their roots through the roughest and most impenetrable rocks and clefts of stone it self; and others require not any rich or pinguid, but very moderate soil; especially, if committed to it in seeds, which allies them to their mother and nurse without reniency or regret: And then considering what assistances a little care in easing and stirring of the ground about them for a few years does afford them: What cannot a strong plow, a winter mellowing, and summer heats, incorporated with the pregnant turf, or a slight assistance of lime, loam, sand, rotten compost, discreetly mixed (as the case may require) perform even in the most unnatural and obstinate soil? And in such places where anciently woods have grown, but are now unkind to them, the fault is to be reformed by this care; and chiefly, by a sedulous extirpation of the old remainders of roots, and latent stumps, which by their mustiness, and other pernicious qualities, sowre the ground, and poyson the conception; and here-with let me put in this note, that even an over-rich, and pinguid composition, is by no means the proper bed either for seminary or nursery, whilst even the natural soil itself does frequently discover and point best to the particular species, though some are for all places alike: Nor should the earth be yet perpetually crop'd with the same, or other seeds, without due repose, but lie some time fallow to receive the influence of heaven, according to good husbandry. But I shall say no more of these particulars at this time, because the rest is sprinkl'd over this whole work in their due places; wherefore we hasten to the following title; namely, the choice and ordering of the seeds.

3. Chuse your seed of that which is perfectly mature, ponderous and sound; commonly that which is easily shaken from the boughs, or gathered about November, immediately upon its spontaneous fall, or taken from the tops and summities of the fairest and soundest trees, is best, and does (for the most part) direct to the proper season of interring, &c. according to institution.

Nature herself who all created first,
Invented sowing, and the wild plants nurs't:
When mast and berries from the trees did drop,
Succeeded under by a numerous crop.[1]

Yet this is to be consider'd that if the place you sow in be too cold for an autumnal semination, your acorns, mast, and other seeds may be prepared for the vernal by being barrel'd, or potted up in moist sand, or earth stratum s.s. during the winter; at the expiration whereof you will find them sprouted; and being committed to the earth, with a tender hand, as apt to take as if they had been sown with the most early; nay, with great advantage: By this means too, they have escaped the vermine, (which are prodigious devourers of winter-sowing) and will not be much concern'd with the increasing heat of the season, as such as being crude, and unfermented, are newly sown in the beginning of the spring; especially, in hot and loose grounds; being already in so fair a progress by this artificial preparation; and which, (if the provision to be made be very great) may be thus manag'd. Chuse a fit piece of ground, and with boards (if it have not that position of it self) design it three foot high; lay the first foot in fine earth, another of seeds, acorns, mast, keys, nuts, haws, holly-berries , &c. promiscuously, or separate, with (now and then) a little mould sprinkled amongst them: The third foot wholly earth: Of these preparatory magazines make as many, and as much larger ones as will serve your turn, continuing it from time to time as your store is brought in. The same for ruder handlings, may you also do by burying your seeds in dry sand, or pulveriz'd earth, barrelling them (as I said) in tubs, or laid in heaps in some deep cellar where the rigour of the winter may least prejudice them; and I have fill'd old hampers, bee-hives, and boxes with them, and found the like advantage, which is to have them ready for your seminary, as before hath been shew'd, and exceedingly prevent the season. There be also who affirm, that the careful cracking and opening of stones which include the kernels, as soon as ripe, precipitate growth, and gain a years advance; but this is erroneous. Now if you gather them in moist weather, lay them a drying, and so keep them till you sow, which may be as soon as you please after Christmas. If they spire out before you sow them, be sure to commit them to the earth before the sprout grows dry.

Note

1   Nam specimen sationis, & infitionis origo
     Ipsa fuit rerum primum natura creatrix:
     Arboribus quoniam baccae, glandesque caducae
     Tempestiva dabant pullorum examina subter, &c.
                                                    (Lucretius, I, 5)

Sylva, London (1678), 2–5, 255–60.

Percy Bysshe Shelley

A vindication of natural diet

In proportion to the number of proselytes, so will be the weight of evidence; and when a thousand persons can be produced, living on vegetables and distilled water, who have to dread no disease but old age, the world will be compelled to regard animal flesh and fermented liquors as slow but certain poison. The change which would be produced by simpler habits on political economy is sufficiently remarkable. The monopolising eater of animal flesh would no longer destroy his constitution by devouring an acre at a meal, and many loaves of bread would cease to contribute to gout, madness, and apoplexy, in the shape of a pint of porter or a dram of gin, when appeasing the long-protracted famine of the hard-working peasant's hungry babes. The quantity of nutritious vegetable matter consumed in fattening the carcase of an ox, would afford ten times the sustenance, undepraving indeed, and incapable of generating disease, if gathered immediately from the bosom of the earth.

The most fertile districts of the habitable globe are now actually cultivated by men for animals, at a delay and waste of aliment absolutely incapable of calculation. It is only the wealthy that can, to any great degree, even now, indulge the unnatural craving for dead flesh, and they pay for the greater licence of the privilege, by subjection to supernumerary diseases. Again, the spirit of the nation that should take the lead in this great reform would insensibly become agricultural: commerce, with all its vice, selfishness, and corruption, would gradually decline; more natural habits would produce gentler manners, and the excessive complication of political relations would be so far simplified that every individual might feel and understand why he loved his country, and took a personal interest in its welfare. How would England, for example, depend on the caprices of foreign rulers, if she contained within herself all the necessaries, and despised whatever they possessed of the luxuries of life? How could they starve her into compliance with their views? Of what consequence would it be that they refused to take her woollen manufactures, when large and fertile tracts of the island ceased to be allotted to the waste of pasturage? On a natural system of diet, we should require no spices from India; no wines from Portugal, Spain, France, or Madeira; none of those multitudinous articles of luxury, for which every corner of the globe is rifled, and which are the causes of so much individual rivalship, such calamitous and sanguinary national disputes.

In the history of modern times, the avarice of commercial monopoly, no less than the ambition of weak and wicked chiefs, seems to have fomented the universal discord, to have added stubbornness to the mistakes of cabinets, and indocility to the infatuation of the people. Let it ever be remembered, that it is the direct influence of commerce to make the interval between the richest and the poorest man wider and more unconquerable. Let it be remembered that it is a foe to every thing of real worth and excellence in the human character. The odious and disgusting aristocracy of wealth, is built upon the ruins of all that is good in chivalry or republicanism; and luxury is the forerunner of a barbarism scarce capable of cure. Is it impossible to realize a state of society, where all the energies of man shall be directed to the production of his solid happiness?

Certainly, if this advantage (the object of all political speculation) be in any degree attainable, it is attainable only by a community which holds out no factitious incentives to the avarice and ambition of the few, and which is internally organized for the liberty, security, and comfort of the many. None must be entrusted with power (and money is the completest species of power) who do not stand pledged to use it exclusively for the general benefit. But the use of animal flesh and fermented liquors, directly militates with this equality of the rights of man. The peasant cannot gratify these fashionable cravings without leaving his family to starve. Without disease and war, those sweeping curtailers of population, pasturage would include a waste too great to be afforded. The labour requisite to support a family is far lighter[1] than is usually supposed. The peasantry work, not only for themselves, but for the aristocracy, the army, and the manufacturers.

The advantage of a reform in diet is obviously greater than that of any other. It strikes at the root of the evil. To remedy the abuses of legislation, before we annihilate the propensities by which they are produced, is to suppose, that by taking away the effect, the cause will cease to operate. But the efficacy of this system depends entirely on the proselytism of individuals, and grounds its merits, as a benefit to the community, upon the total change of the dietetic habits in its members. It proceeds securely from a number of particular cases to one that is universal, and has this advantage over the contrary mode, that one error does not invalidate all that has gone before.

Note

1   It has come under the author's experience that some of the workmen on an embankment in North Wales who, in consequence of the inability of the proprietor to pay them, seldom received their wages, have supported large families by cultivating small spots of sterile ground by moonlight. In the notes to Pratt's Poem, 'Bread for the Poor,' is an account of an industrious labourer, who by working in a small garden, before the after his day's task, attained to an enviable state of independence.

A Vindication of Natural Diet (1813), London: Vegetarian Society, 20–3.

François Fourier

The economics of harmony

We shall find among the Harmonians a policy totally contrary to our ideas of commerce, which promote waste and the changes of fashion, under the pretext of maintaining the workman. But in Harmony the workman, the agriculturist, and the consumer, are one and the same person; he has no interest in practising extortion upon himself, as in civilisation, where everyone strives to promote industrial disturbance occasioned by changes of fashion, and to manufacture poor goods or poor furniture, in order to double consumption, to enrich the merchants at the expense of the people and of real wealth.

They will calculate, in Harmony, that changes of fashion, defective quality, or imperfect workmanship, would cause a loss of five hundred francs per individual, for the poorest of the Harmonians possesses a wardrobe of clothes for all seasons, and is accustomed to using furniture, trappings, and appurtenances, for work or pleasure, of a fine quality.

They do not calculate thus in civilisation, because that society, in industry as in everything, is inclined to duplicity or internal warfare. Its industry is a veritable civil war of the producer against the idler, whom he tries to plunge into ruin; and of the merchant against the social body, which he incites to dupery. The science which applauds this conflict resembles a senseless master who should incite his domestics to break quantities of dishes and furniture, for the benefit of the manufacturers. Everything is but political madness, as long as the interest of the individual is not bound up with the interest of the mass.

Let us refute a strange sophism of the economists who maintain that the unlimited increase of manufactured products is an increase of wealth; the consequence of that would be, that if every person could be induced to use four times as many clothes as he does, the social world would attain to four times its present wealth in manufactured products.

No truth whatever in this; their calculation is as false on this point as it is on the desirability of unlimited increase of population, or *food for cannon*. Real wealth, in Harmony, is based upon:

The greatest possible consumption of varieties of food;

The smallest possible consumption of varieties of clothing and furniture.

Variety, applied to both kinds of consumption, demands the maximum on one side and the minimum on the other, all harmony being based upon direct and inverse action of impulses.

This principle has escaped civilised economists, who, likening manufactures to agriculture, have believed that excessive manufacturing and consumption of goods is a measure of the increase of wealth. The speculations of Harmony upon this point are the reverse; it desires, in clothing and in furniture, *infinite variety*, but the *smallest consumption*.

When I was little practised in the calculations of attraction, and I began to balance the portions and the results, in every branch of industry, I was greatly astonished to find that, strictly analysed, there was little attraction for manufacturing labour, and that the associative order, while creating agricultural allurements in unlimited quantities, would develop only an insignificant amount of manufacturing allurements. This result seemed inconsistent to me, opposed to what necessity demanded. Little by little, I perceived that, in accordance with the principle of attractions proportioned to ends, God ought to have restricted the allurements of manufacture, by reason of the excellence of associative industry, which raises every manufactured article to the acme of perfection, so that furniture and clothing attain prodigious durability, become *everlasting*.

Shoes made by a fashionable shoemaker of Paris will go into holes without fail after a month's wear; and this is as it should be; for that shoemaker would compromise his art, *if he should furnish common people who go about on foot*. The shoes coming from the workshops of a Phalanx will be in good condition at the end of ten years, because two conditions, unknown in the present order, will have been fulfilled:

Excellence of material and of workmanship;

Fitness for its purpose and for durability.

These details, sordid in appearance, become sublime when one considers that they are capable of securing an *annual* saving of 400 milliards in wearing apparel, and 2,000 milliards in the total less which would be incurred if the Harmonians failed to take combined saving into their calculations.

With them, economy becomes *bon ton*, through the influence of the combined action of the four tones. The Harmonians, though liberal and fond of elegance, have a passion, as being *bon ton*, for savings which we regard as niggardliness, sordid avarice, such as the picking up of a pin or using the other end of a match. They will treat you profusely to the finest dishes, and regard you as a vandal if you waste a cherry-seed or the skin of an apple.

With us, as a matter of propriety, one writes to the minister upon paper of ample dimensions, three-fourths of which is useless, and the minister, by way of fiscal parade, replies with two lines upon a sheet a yard long. A contrary spirit will prevail among the Harmonians, and, in writing to the minister, honesty will demand that the smallest quantity of paper possible shall be used. To fail to do that would be to offend the minister, to suppose him indifferent to small savings, which in Harmony are the guarantee of social happiness, not only on account of the annual gain of 2,000 milliards, but on account of maintaining the equilibrium between services and attractions. This equilibrium would be destroyed, if an excessive consumption of manufactured articles were to divert people from the pleasant agricultural sessions, and oblige them to take hours from such labour and devote them to manufacture, whose allurements are limited in quantity, while agricultural Attraction is unlimited.

In an order in which all classes will be linked by ties of affection, potentates themselves will be found to set the fashion in that economy of clothes which we characterise as a sordid spirit, and which is the real spirit of God, whose first quality is the economy of means. God does not waste an atom in the mechanism of the universe, and everywhere where there is an absence of general economy, we may say there is an absence of the spirit of God.

'The economics of harmony' (1822), *Selections from the Works of Fourier*, New York: Gordon (1972), 196–8.

G. P. Marsh

Recycling and destruction

Man has too long forgotten that the earth was given to him for usufruct alone, not for consumption, still less for profligate waste. Nature has provided against the absolute destruction of any of her elementary matter, the raw material of her works; the thunder-bolt and the tornado, the most convulsive throes of even the volcano and the earthquake, being only phenomena of decomposition and recomposition. But she has left it within the power of man irreparable to derange the combinations of inorganic matter and of organic life, which through the night of aeons she had been proportioning and balancing, to prepare the earth for his habitation, when, in the fulness of time, his Creator should call him forth to enter into its possession.

Apart from the hostile influence of man, the organic and the inorganic world are, as I have remarked, bound together by such mutual relations and adaptations as secure, if not the absolute permanence and equilibrium of both, a long continuance of the established conditions of each at any given time and place, or at least, a very slow and gradual succession of changes in those conditions. But man is everywhere a disturbing agent. Wherever he plants his foot, the harmonies of nature are turned to discords. The proportions and accommodations which insured the stability of existing arrangements are overthrown. Indigenous vegetable and animal species are extirpated, and supplanted by others of foreign origin, spontaneous production is forbidden or restricted, and the face of the earth is either laid bare or covered with a new and reluctant growth of vegetable forms, and with alien tribes of animal life. These intentional changes and substitutions constitute, indeed, great revolutions; but vast as is their magnitude and importance, they are, as we shall see, insignificant in comparison with the contingent and unsought results which have flowed from them.

The fact that, of all organic beings, man alone is to be regarded as essentially a destructive power, and that he wields energies to resist which, nature – that nature whom all material life and all inorganic substance obey – is wholly impotent, tends to prove that, though living in physical nature, he is not of her, that he is of more exalted parentage, and belongs to a higher order of existences than those born of her womb and submissive to her dictates.

There are, indeed, brute destroyers, beasts and birds and insects of prey – all animal life feeds upon, and, of course, destroys other life – but this destruction is balanced by compensations. It is, in fact, the very means by which the existence of one tribe of animals or of vegetables is secured against being smothered by the encroachments of another; and the reproductive powers of species, which serve as the food of others, are always proportioned to the demand they are destined to supply. Man pursues his victims with reckless destructiveness; and, while the sacrifice of life by the lower animals is limited by the cravings of appetite, be unsparingly persecutes, even to extirpation, thousands of organic forms which he cannot consume.[35]. . .

But man, the domestic animals that serve him, the field and garden plants the products of which supply him with food and clothing, cannot subsist and rise to the full development of their higher properties, unless brute and unconscious nature be effectually combated, and, in a great degree, vanquished by human art. Hence, a certain measure of transformation of terrestrial surface, of suppression of natural, and stimulation of artificially modified productivity becomes necessary. This measure man has unfortunately exceeded. He has felled the forests whose network of fibrous roots bound the mould to the rocky skeleton of the earth; but had he allowed here and there a belt of woodland to reproduce itself by spontaneous propagation, most of the mischiefs which his reckless destruction of the natural protection of the soil has occasioned would have been averted. He has broken up the mountain reservoirs, the percolation of whose waters through unseen channels supplied the fountains that refreshed his cattle and fertilized his fields; but he has neglected to maintain the cisterns and the canals of irrigation which a wise antiquity had constructed to neutralize the consequences of its own imprudence. While he has torn the thin glebe which confined the light earth of extensive plains, and has destroyed the fringe of semi-aquatic plants which skirted the coast and checked the drifting of the sea sand, he has failed to prevent the spreading of the dunes by clothing them with artificially propagated vegetation. He has ruthlessly warred on all the tribes of animated nature whose spoil he could convert to his own uses, and he has not protected the birds which prey on the insects most destructive to his own harvests.

Note

35 The terrible destructiveness of man is remarkably exemplified in the chase of large mammalia and birds for single products, attended with the entire waste of enormous quantities of flesh, and of other parts of the animal, which are capable of valuable uses. The wild cattle of South America are slaughtered by millions for their hides and horns; the buffalo of North America for his skin or his tongue; the elephant, the walrus, and the narwhal for their tusks; the cetacea, and some other marine animals, for their oil and whalebone; the ostrich and other large birds, for their plumage. Within a few years, sheep have been killed in New England by whole flocks, for their pelts and suet alone, the flesh being thrown away; and it is even said that the bodies of the same quadrupeds have been used in Australia as fuel for limekilns. What a vast amount of human nutriment, of bone, and of other animal products valuable in the arts, is thus recklessly squandered! In nearly all these cases, the part which constitutes the motive for this wholesale destruction, and is alone saved, is essentially of insignificant value as compared with what is thrown away. The horns and hide of an ox are not economically worth a tenth part as much as the entire carcass.

One of the greatest benefits to be expected from the improvements of civilization is, that increased facilities of communication will render it possible to transport to places of consumption much valuable material that is now wasted because the price at the nearest market will not pay freight. The cattle slaughtered in South America for their hides would feed millions of the starving population of the Old World, if their flesh could be economically preserved and transported across the ocean.

We are beginning to learn a better economy in dealing with the inorganic world. The utilization – or, as the Germans more happily call it, the Verwerthung, the *beworthing* – of waste from metallurgical, chemical, and manufacturing establishments, is among the most important results of the application of science to industrial purposes. The incidental products from the laboratories of manufacturing chemists often become more valuable than those for the preparation of which they were erected. The slags from silver refineries, and even from smelting houses of the coarser metals, have not unfrequently yielded to a second operator a better return than the first had derived from dealing with the natural ore; and the saving of lead carried off in the smoke of furnaces has, of itself, given a large profit on the capital invested in the works. A few years ago, an officer of an American mint was charged with embezzling gold committed to him for coinage. He insisted, in his defence, that much of the metal was volatilized and lost in refining and melting, and upon scraping the chimneys of the melting furnaces and the roofs of the adjacent houses, gold enough was found in the soot to account for no small part of the deficiency.

Man and Nature (1874), Cambridge, Mass.: Belknap (1965), 36–8.

Gifford Pinchot

Conservation

The central thing for which Conservation stands is to make this country the best possible place to live in, both for us and our descendants. It stands against the waste of natural resources which cannot be renewed, such as coal and iron; it stands for the perpetuation of the resources which can be renewed, such as the food-producing soils and the forests; and most of all it stands for an equal opportunity for every American citizen to get his fair share of benefit from these resources, both now and hereafter.

Conservation stands for the same kind of practical commonsense management of this country by the people that every businessman stands for in the handling of his own business. It believes in prudence and foresight instead of reckless blindness; it holds that resources now public property should not become the basis for oppressive private monopoly; and it demands the complete and orderly development of all our resources for the benefit of all the people instead of the partial exploitation of them for the benefit of a few. It recognises fully the right of the present generation to use what it needs and all it needs of the natural resources now available, but it recognises equally our obligation so to use what we need that our descendants shall not be deprived of what they need.

Conservation has much to do with the welfare of the average man of today. It proposes to secure a continuous and abundant supply of the necessaries of life, which means a reasonable cost of living and business stability. It advocates fairness in the distribution of the benefits which flow from the natural resources.

The Fight for Conservation, Garden City, N.Y.: Harcourt Brace (1901), 79–81.

CHAPTER 11

The Frankenstein factor

J. Needham, Taoism and science

M. Shelley, Frankenstein

T. Carlyle, Signs of the times

T. Carlyle, To Alexander Carlyle

*G. C. Burrows, A word to the electors on the
unrestricted use of modern machinery*

K. Marx, A crippled monstrosity

E. Carpenter, Civilization: its cause and cure

G. Orwell, Science and socialism

Introduction

> Certain tools are destructive no matter who owns them, whether it be the
> Mafia, stockholders, a foreign company, the state or even a workers' commune.
>
> (Illich 1975: 39)

Ironically while scientific research is marshalled by Greens to further their arguments (witness the use of scientific evidence to discredit nuclear power or to draw attention to the greenhouse effect) the movement draws upon a long tradition critical of science and technology. Suspicion of the 'technofix' is common, with nuclear power, the chemical industry and biotechnology viewed with particular hostility. Such technophobia derives from a number of distinct yet linked concerns. First, as Capra (1983) argues, a reductionist scientific method is innately flawed and hostile to nature. While science may not be rejected (such critics tend to be keen on ecology and the New Physics), what is usually seen as orthodox scientific method is criticized. A new holistic science is called for, which draws on and in turn enhances ecological appreciation. Second, modern science, far from being defective, may be seen as all too effective at revealing nature's secrets. Greens fear that effective technology allows humanity to rape nature more effectively. Science is seen as being out of control, with scientists developing new and potentially dangerous technologies as a means of expanding their control over others or simply for the hell of it, out of curiosity. Greens, especially those of a deep ecology or leftist orientation, suspect that inappropriate and atavistic techno-fixes are proposed as the solution of social or environmental problems that demand democratic and political answers.

In our first passage, from Needham's authoritative study of science in ancient China, we see that such fears are far from new. The Taoist sage spurns the emperor's demand that he should tell him how to manipulate nature for his benefit. In a prophecy of the misfortunes of misapplied technology Tzu fears that the emperor, if informed of the true workings of the Tao, would wreak havoc: 'the herbs and trees would shed their leaves . . . the light of the sun and moon would hasten to extinction'. Taoist feeling for nature, its anarchism and commitment to a communist society which balances spiritual growth with human freedom, is certainly worthy of study by modern Greens. The Taoist relationship with earlier shamanistic pantheism is an area that demands careful research. Needham (1956: 2, 98) notes the similarity of the Taoist outlook to the peasant communism of the seventeenth-century English Diggers, whom we will be hearing from later (Chapter Seventeen).

We have already noted the prefigurative Green beliefs inherent in the Romantic movement of Blake, Goethe and Shelley. Mary Shelley (1797–1851), daughter of the feminist Mary Wollstonecraft, the anarchist Godwin and wife of Percy Bysshe, is, after Blake, perhaps the most important of this group, in a modern Green context. With *Frankenstein* she provides the story of a technologist who creates artificial life which becomes uncontrollable and eventually destroys its supposed master. She may also be described as the first science-fiction author and as such a potent source of technological misgivings. Here Dr Frankenstein draws parallels between his work and that of Newton, a target, as we have seen, for both Blake and Capra.

Carlyle (1795–1881), a conservative second-generation Victorian Romantic, bitterly attacks the 'machine age' in his essay *Signs of the Times* and in a letter to his brother, Alexander, notes the growing effects of polluting technology. Burrows, his contemporary, a leftist – one might even say, a Luddite – critic of technology, warns that machines control individuals. In a short passage from *Capital* we find the far from technophobic Marx describing the harm done to the human body and soul by the factory system and, in particular, the division of labour. Both Marx and Burrows fear that 'mankind are slaves to things inanimate'. Greens, on the contrary, go beyond this concern, suspecting that these 'things inanimate' are tools used by humanity to make nature a slave.

Edward Carpenter (1844–1929), whose varied writings provide the most exact nineteenth-century equivalent of late twentieth-century Green thinking, puts concisely the case for criticism. He is not anti-technology but is cautious and wishes to see science placed in an environmentally sound and human-centred context. George Orwell (1903–1950), the English socialist who had little time for 'vegetarians and fruit juice drinkers' (a clear reference to the likes of Carpenter), was a pragmatist inspired by 'common decency'. No utopian, he was clearly worried that the socialism of the 1930s was too closely associated with an alien industrial vision. Why socialism in Britain, the United States and Russia had so swiftly moved within a generation from a rural, rather bucolic vision to one that sought a high-technology society is a fascinating question that cannot be explored here. Orwell's account contrasts with the anti-industrial radicalism advocated by Morris, Carpenter, the Guild Socialists, Narodniks and even some Marxists prior to the 1920s. The extract from *The Road to Wigan Pier* is a fine piece of writing that explores the contradictions of industrialism. It also provides a survey of the wider literature dealing with society, applied science and nature.

J. Needham

Taoism and science

This contrast is detectable in certain passages of Chuang Tzu which almost take a form reminiscent of modern discussions about science and social welfare. These parables and imaginary conversations seem surely intended to imply that the application of science to human benefit was premature, and that what the Confucians should do if they really wanted to apply human knowledge for the improvement of the conditions of the life of Man was to become Taoists and devote themselves first to the observation of Nature. To help Man without understanding Nature was impossible. Thus, chapter 11:

> Huang Ti[1] had been on the throne for nineteen years, and his writ was running everywhere in the empire, when he heard that Kuang Chhêng Tzu[2] was living on the top of Mount Empty-togetherness, so he went there to see him.
>
> 'I have heard,' he said, 'that you, Sir, are profoundly learned in the perfect Tao. May I ask what is its essence? I wish to take the subtlest essences of heaven and earth and assist with them the (growth of the) five cereal grains, for the (better) nourishment of the people. I also wish to direct the (operations of the) Yin and the Yang, so as to secure the comfort of all living beings. How should I proceed?'
>
> Kuang Chhêng Tzu replied, 'What you are asking about is the material basis of things (*wu chih chih yeh*); what you desire to control can only be the scattered fragments of these things (*wu chih tshan yeh*) (which have been destroyed by your previous interference). According to your government of the world, the vapours of the clouds, before they were collected, would descend in rain; the herbs and trees would shed their leaves before they became yellow; and the light of the sun and moon would hasten to extinction. You have the shallow mind of a glib talker; it is not fit that I should tell you about the perfect Tao.'

Kuang Chhêng Tzu reproaches Huang Ti for the superficial approach to Nature whereby immediate advantages are sought from the broken fragments of the material manifestations of things. He hints that the only way really to benefit human society is to go back and elucidate the fundamental principles of Nature. Huang Ti's attitude is compared to that of a greedy plunderer of Nature, who would allow neither clouds nor crops to ripen, instead of waiting to find out and apply the basic principles of Nature. Bearing in mind what mankind knows today about soil conservation and nature protection, and all the experience we have gained as to the proper relations between pure and applied science, this passage of Chuang Tzu seems as profound and prophetic as any he ever wrote.

Notes

1 Legendary emperor.
2 Imaginary hermit.

Science and Civilisation in China, Cambridge: Cambridge University Press (1956), 2: 98–9.

Mary Wollstonecraft Shelley

Frankenstein, or, The Modern Prometheus

I have described myself as always having been embued with a fervent longing to penetrate the secrets of nature. In spite of my intense labour and wonderful discoveries of modern philosophers, I always came from my studies discontented and unsatisfied. Sir Isaac Newton is said to have avowed that he felt like a child picking up shells beside the great and unexplored ocean of truth. Those of his successors in each branch of natural philosophy with whom I was acquainted appeared, even to my boy's apprehensions, as tyros engaged in the same pursuit. . . .

The untaught peasant beheld the elements around him, and was acquainted with their practical uses. The most learned philosopher knew little more. He had partially unveiled the face of Nature, but her immortal lineaments were still a wonder and a mystery. He might dissect, anatomise, and give names; but, not to speak of a final cause, causes in their secondary and tertiary grades were utterly unknown to him. I had gazed upon the fortifications and impediments of nature, and rashly and ignorantly I had repined. . . .

By one of those caprices of the mind, which we are perhaps most subject to in early youth, I at once gave up my former occupations; set down natural history and all its progeny as a deformed and abortive creation; and entertained the greatest disdain for a would-be science, which could never even step within the threshold of real knowledge. In this mood of mind I betook myself to the mathematics, and the branches of study appertaining to that science, as being built upon secure foundations, and so worthy of my consideration.

Thus strangely are our souls constructed, and by such slight ligaments are we bound to prosperity or ruin. When I look back, it seems to me as if this almost miraculous change of inclination and will was the immediate suggestion of the guardian angel of my life – the last effort made by the spirit of preservation to avert the storm that was even then hanging in the stars, and ready to envelope me . . . It was a strong effort of the spirit of good; but it was ineffectual. Destiny was too potent, and her immutable laws had decreed my utter and terrible destruction. . . .

Partly from curiosity, and partly from idleness, I went into the lecturing room, which M. Waldman entered shortly after. This professor was very unlike his colleague. He appeared about fifty years of age, but with an aspect expressive of the greatest benevolence; a few grey hairs covered his temples, but those at the back of his head were nearly black. His person was short, but remarkably erect; and his voice the sweetest I have ever heard. He began his lecture by a recapitulation of the history of chemistry, and the various improvements made by different men of learning, pronouncing with fervour the names of the most distinguished discoverers. He then took a cursory view of the present state of the science, and explained many elementary terms. After having made a few preparatory experiments, he concluded with a panegyric upon modern chemistry, the terms of which I shall never forget:

'The ancient teachers of this science,' said he, 'promised impossibilities, and performed nothing. The modern masters promised very little; they know that metals cannot be transmuted, and that the elixir of life is a chimera. But these philosophers, whose hands seem only made to dabble in dirt, have indeed performed miracles. They penetrate into the recesses of nature, and show how she works in her hiding places. They ascend into the heavens: they have discovered how the blood circulates, and the nature of the air we breathe. They have acquired new and almost unlimited powers; they can command the thunders of heaven, mimic the earthquakes, and even mock the invisible world with its own shadows.'

Such were the professor's words – rather let me say such the words of fate, enounced to destroy me. As he went on, I felt as if my soul were grappling with a palpable enemy; one by one the various keys were touched which formed the mechanism of my being: chord after chord was sounded, and soon my mind was filled with one thought, one conception, one purpose. So much has been done, exclaimed the soul of Frankenstein – more, far more, will I achieve: treading in the steps already marked, I will pioneer a new way, explore unknown powers, and unfold to the world the deepest mysteries of creation....

I then thought my father would be unjust if he ascribed my neglect to vice, or faultiness on my part; but I am now convinced that he was justified in conceiving that I should not altogether be free from blame. A human being in perfection ought always to preserve a calm and peaceful mind, and never allow passion or a transitory desire to disturb his tranquillity. I do not think that the pursuit of knowledge is an exception to this rule. If the study to which you apply yourself has a tendency to weaken your affections, and to destroy your taste for those simple pleasures in which no alloy can possible mix, then that study is certainly unlawful, that is to say, not befitting the human mind. If this rule were always observed; if no man allowed any pursuit whatsoever to interfere with the tranquillity of his domestic affections, Greece had not been enslaved; Caesar would have spared his country; America would have been discovered more gradually; and the empires of Mexico and Peru had not been destroyed.

But I forgot that I am moralising in the most interesting part of my tale; and your looks remind me to proceed....

'Devil,' I exclaimed, 'do you dare approach me? and do not you fear the fierce vengeance of my arm wreaked on your miserable head? Begone, vile insect! or rather stay, that I may trample you to dust! and, oh! that I could with the extinction of your miserable existence, restore those victims whom you have so diabolically murdered!'

'I expected this reception,' said the daemon. 'All men hate the wretched; how, then, must I be hated, who am miserable beyond all living things! Yet you, my creator, detest and spurn me, thy creature, to whom thou art bound by ties only dissoluble by the annihilation of one of us. You purpose to kill me. How dare you sport with life? Do your duty towards me, and I will do mine towards you and the rest of mankind. If you comply with my conditions, I will leave them and you at peace; but if you refuse, I will glut the maw of death, until it be satiated with the blood of your remaining friends.'

'Abhorred monster! fiend that thou art! the tortures of hell are too mild a vengeance for thy crimes. Wretched devil! you reproach me with your creation; come on, then, that I may extinguish the spark which I so negligently bestowed.'

Frankenstein (1831), London: Oxford University Press (1969), 39, 40, 41–2, 47–8, 55–6, 99.

Thomas Carlyle

Signs of the times

Were we required to characterise this age of ours by any single epithet, we should be tempted to call it, not an Heroical, Devotional, Philosophical, or Moral Age, but, above all others, the Mechanical Age. It is the Age of Machinery, in every outward and inward sense of that word; the age with its whole undivided might forwards, teaches and practises the great art of adapting means to ends. Nothing is now done directly, or by hand; all is by rule and calculated contrivance. For the simplest operation, some helps and accompaniments, some cunning abbreviating process is in readiness. Our old modes of exertion are all discredited, and thrown aside. On every hand, the living artisan is driven from his workshop, to make room for a speedier, inanimate one. The shuttle drops from the fingers of the weaver, and falls into iron fingers that ply it faster. The sailor furls his sail, and lays down his oar; and bids a strong, unwearied servant, in vaporous winds, bear him through the waters. Men have crossed oceans by steam; the Birmingham Fire-king has visited the fabulous East; and the genius of the Cape, were there any Camoens now to sing it, has again been alarmed, and with far stranger thunders than Gamas. There is no end to machinery. Even the horse is stripped of his harness, and finds a fleet fire-horse yoked in his stead. Nay, we have an artist that hatches chickens by steam; the very brood-hen is to be superseded! For all earthly, and for some unearthly purposes, we have machines and mechanic furtherances; for mincing our cabbages; for casting us into magnetic sleep. We remove mountains, and make seas our smooth highway; nothing can resist us. We war with rude Nature; and, by our resistless engines, come off always victorious; and loaded with spoils.

What wonderful accessions have thus been made, and are still making, to the physical power of mankind; how much better fed, clothed, lodged and, in all outward respects, accommodated men now are, or might be, by a given quantity of labour, is a grateful reflection which forces itself on every one. What changes, too, this addition of power is introducing into the Social System; how wealth has more and more increased, and at the same time gathered itself more and more into masses, strangely altering the old relations, and increasing the distance between the rich and the poor, will be a question for Political Economists, and a much more complex and important one than any they have yet engaged with.

The Works of Thomas Carlyle in Thirty Volumes (1829), London: Chapman & Hall, 2: 59–60

Thomas Carlyle

To Alexander Carlyle

... I was one day thro the iron and coal works of this neighbourhood – a half-frightful scene! A space perhaps thirty square miles to the north of us, covered over with furnaces, rolling-mills, steam-engines and sooty men. A dense cloud of pestilential smoke hangs over it for ever, blackening even the grain that grows upon it; and at night the whole region burns like a volcano spitting fire from a thousand tubes of brick. But oh the wretched hundred and fifty thousand mortals that grind out their destiny there! In the coal-mines they were literally naked, any of them, all but trowsers; black as ravens; plashing about among dripping caverns, or scrambling amid heaps of broken mineral; and thirsting unquenchably for beer. In the iron-mills it was little better: blast-furnaces were roaring like the voice of many whirlwinds all around; the fiery metal was hissing thro' its moulds, or sparkling and spitting under hammers of a monstrous size, which fell like so many little earthquakes. Here they were wheeling charred coals, breaking their iron-stone, and tumbling all into their fiery pit; there they were turning and boring cannon with a hideous shrieking noise such as the earth could hardly parallel; and thro' the whole, half-naked demons pouring with sweat and besmeared with soot were hurrying to and fro in their red nightcaps and sheet-iron breeches rolling or hammering or squeezing their glowing metal as if it had been wax or dough. They also had a thirst for ale. Yet on the whole I am told they are very happy: they make forty shillings or more per week, and few of them will work on Mondays. It is in a spot like this that one sees the sources of British power. The skill of man combining these coals and that iron-ore (till forty years ago – iron was smelted with charcoal only) has gathered three or four hundred thousand human beings round this spot, who send the products of their industry to all the ends of the Earth.

'Letter to Alexander Carlyle, 11th August, 1924', in *The Collected Letters of Thomas and James Welsh Carlyle*, Durham, NC: Duke University Press (1970), 125–6.

G. C. Burrows

A word to the electors on the unrestricted use of modern machinery

Machinery is the hydra of the present day, starvation is her offspring, and as long as the land is cursed with unrestricted machinery, machinery vying with itself, the inhabitants of the whole earth cannot consume the produce. Every market must be glutted, the industry of the human race be of no avail ... nations compete with nations, machinery with machinery; to get this gold, mankind are slaves to things inanimate ...

A Word to the Electors on the unrestricted Use of modern Machinery, Norwich (1832).

Karl Marx

A crippled monstrosity

It converts the worker into a crippled monstrosity by furthering his particular skill as in a forcing-house, through the suppression of a whole world of productive drives and inclination, just as in La Plata, they butcher a whole beast for the sake of his hide or his tallow. Not only is specialised work distributed among the different individuals, but the individual himself is divided up, and transformed into the automatic motor of a detail operation, thus realising the absurd fable of Menenuis Agrippa, which presents man as a mere fragment of his own body. If, in the first place, the worker sold his labour-power to capital because he lacked the material means of producing a commodity, now his own individual labour-power withholds its services unless it has been sold to capital. It will function only in an environment which first comes into existence after its sale, namely the capitalist's workshop. Unfitted by nature to make anything independently, the manufacturing worker develops his productive activity only as an appendage of that workshop. As the chosen people bore in their features the sign that they were the property of Jehovah, so the division of labour brands the manufacturing worker as the property of capital.

Capital (1867), Harmondsworth: Penguin (1979), 1: 481–2.

Edward Carpenter

Civilization: its cause and cure

In 1889 I got off *Civilization: Its Cause and Cure* – another series of reprints. And here too the philosophical position, though often crudely expressed, and with more attempt at *suggestion* than finish, is I think in the main well founded and valuable. The attacks on Civilization and on Modern Science were both wrung from me, as it were, by some inner evolution or conviction and against my will; but in both cases the position once taken became to me fully justified. In neither case did I take any great precautions to guard against misunderstanding, and in consequence I have been freely accused of blinding myself – in respect of Civilization – to modern progress, and of desiring to return to the state of primitive man; and in respect to Science – of preferring ignorance to intelligence. But no careful reader would make these mistakes. The monumental, patient, one may almost say heroic, work which has been done by Science during the nineteenth century, in the way of exact observation, classification, and detailed practical application, can never be ignored and can hardly be over-estimated. None the less the very decided criticism in *Civilization: Its Cause and Cure* of the limits of scientific theorizing and authority has been quite necessary; as well as the forcible insistence on the fact that Science only deals with the surface of life and not with its substance. As to Civilization the advances of Humanity during the Civilization period have been largely bound up with the advance of Science and have chiefly consisted perhaps in increase of technical mastery over Nature and materials. Like every increase of power this has led to greater opportunity of good and greater opportunity of evil. On the moral side, however, we may believe that men's sympathies *have* broadened and widened during the civilization period – so that there is a larger and more general sense of Humanity. On the other hand during this period something of the intensity of the old tribal kinship and community of life has been lost, as well as something of the instinctive kinship of each individual to Nature. It is obvious enough that there can be no *return* to pre-scientific or pre-civilization conditions – though it may be hoped that a later age may combine some of the virtues of the more primitive man with the powers that have been gained during civilization.

My Days and Dreams, London: Allen & Unwin (1916), 141–2.

George Orwell

Science and Socialism

The first thing to notice is that the idea of Socialism is bound up, more or less inextricably, with the idea of machine-production. Socialism is essentially an *urban* creed. It grew up more or less concurrently with industrialism, it has always had its roots in the town proletariat and the town intellectual, and it is doubtful whether it could ever have arisen in any but an industrial society. Granted industrialism, the idea of Socialism presents itself naturally, because private ownership is only tolerable when every individual (or family or other unit) is at least moderately self-supporting; but the effect of industrialism is to make it impossible for anyone to be self-supporting even for a moment. Industrialism, once it rises above a fairly low level, *must* lead to some form of collectivism. Not necessarily to Socialism, of course; conceivably it might lead to the Slave-State of which Fascism is a kind of prophecy. And the converse is also true. Machine-production suggests Socialism, but Socialism as a world-system implies machine-production, because it demands certain things not compatible with a primitive way of life. It demands, for instance, constant intercommunication and exchange of goods between all parts of the earth; it demand some degree of centralised control; it demands an approximately equal standard of life for all human beings and probably a certain uniformity of education. We may take it, therefore, that any world in which Socialism was a reality would be at least as highly mechanized as the United States at this moment, probably much more so. In any case, no Socialist would think of denying this. The Socialist world is always pictured as a completely mechanised, immensely organised world, depending on the machine as the civilisations of antiquity depended on the slave.

So far so good, or so bad. Many, perhaps a majority, of thinking people are not in love with machine-civilisation, but everyone who is not a fool knows that it is a nonsense to talk at this moment about scrapping the machine. But the unfortunate thing is that Socialism, as usually presented, is bound up with the idea of mechanical progress, not merely as a necessary development but as an end in itself, almost as a kind of religion. This idea is implicit in, for instance, most of the propagandist stuff that is written about the rapid mechanical advance in Soviet Russia (the Dneiper dam, tractors, etc., etc.). Karel Capek hits it off well enough in the horrible ending of *R.U.R.*, when the Robots, having slaughtered the last human being, announce their intention to 'build many houses' (just for the sake of building houses, you see). The kind of person who most readily accepts Socialism is also the kind of person who views mechanical progress, *as such*, with enthusiasm. And this is so much the case that Socialists are often unable to grasp that the opposite opinion exists. As a rule the most persuasive argument they can think of is to tell you that the present mechanisation of the world is as nothing to what we shall see when Socialism is established. Where there is one aeroplane now, in those days there will be fifty! All the work that is now done by hand will then be done by machinery: everything that is now made of leather, wood or stone will be made of rubber, glass or steel; there will be no disorder, no loose ends, no wildernesses, no wild animals, no weeds, no disease, no poverty, no pain – and so on and so forth. . . .

The sensitive person's hostility to the machine is in one sense unrealistic, because of the obvious fact that the machine has come to stay. But as an attitude of mind there is a great deal to be said for it. The machine has got to be accepted, but it is probably better to accept it rather as one accepts a drug – that is, grudgingly and suspiciously. Like a drug, the machine is useful, dangerous and habit-forming. The oftener one surrenders to it the tighter its grip becomes. You only have to look about you at this moment to realise with what sinister speed the machine is getting us into its power. To begin with, there is the frightful debauchery of taste that has already been effected by a century of mechanisation. This is almost too obvious and too generally admitted to need pointing out. But as a single instance, take taste in its narrowest sense – the taste for decent food. In the highly mechanised countries, thanks to tinned food, cold storage, synthetic flavouring matters, etc., the palate is almost a dead organ. As you can see by looking at any greengrocer's shop, what the majority of English people mean by an apple is a lump of highly-coloured cotton wool from America or Australia; they will devour these things, apparently with pleasure, and let the English apples rot under the trees. It is the shiny, standardised, machine-made look of the American apple that appeals to them; the superior taste of the English apple is something they simply do not notice. Or look at the factory-made, foil-wrapped cheeses and 'blended' butter in any grocer's; look at the hideous rows of tins which usurp more and more of the space in any food-shop, even a dairy; look at a sixpenny

Swiss roll or a twopenny ice-cream; look at the filthy chemical by-product that people will pour down their throats in under the name of beer. Wherever you look you will see some slick machine-made article triumphing over the old-fashioned article that still tastes of something other than sawdust. And what applies to food applies also to furniture, houses, clothes, books, amusements and everything else that makes up our environment. There are now millions of people, and they are increasing every year, to whom the blaring of a radio is not only a more acceptable but a more *normal* background to their thoughts than the lowing of cattle or the song of birds. The mechanisation of the world could never proceed very far while taste, even the taste-buds of the tongue, remain uncorrupted, because in that case most of the products of the machine would be simply unwanted. In a healthy world there would be no demand for tinned foods, aspirins, gramophones, gaspipe chairs, machine guns, daily newspapers, telephones, motor-cars, etc., etc.; and on the other hand there would be a constant demand for the things the machine cannot produce. But meanwhile the machine is here and its corrupting effects are almost irresistible. One inveighs against it, but one goes on using it. Even a bare-arse savage, given the chance, will learn the vices of civilisation within a few months. Mechanisation leads to the decay of taste, the decay of taste leads to the demand for machine-made articles and hence to more mechanisation, and so a vicious circle is established.

But in addition to this there is a tendency for the mechanisation of the world to proceed as it were automatically, whether we want it or not. This is due to the fact that in modern Western man the faculty of mechanical invention has been fed and stimulated till it has reached almost the status of an instinct. People invent new machines and improve existing ones almost unconsciously, rather as a somnambulist will go on working in his sleep. . . .

. . . In every country in the world the large army of scientists and technicians, with the rest of us panting at their heels, are marching along the road of 'progress' with the blind persistence of a column of ants. Comparatively few people want it to happen, plenty of people actively want it *not* to happen, and yet it is happening. The process of mechanisation has itself become a machine, a huge glittering vehicle whirling at us we are not certain where, but probably towards the padded Wells-world and the brain in the bottle.

This, then, is the case against the machine. Whether it is a sound or unsound case hardly matters. The point is that these or very similar arguments would be echoed by every person who is hostile to machine-civilisation. And unfortunately, because of that nexus of thought, 'Socialism–progress–machinery–Russia–tractors–hygiene–machinery–progress', which exists in almost everyone's mind, it is usually the *same* person who is hostile to Socialism. The kind of person who hates central heating and gaspipe chairs is also the kind of person who, when you mention Socialism, murmurs something about 'beehive state' and moves away with a pained expression. . . . the Socialist will usually assume that you want to revert to a 'state of nature' – meaning some stinking palaeolithic cave: as though there were nothing between a flint scraper and the steel mills of Sheffield, or between a skin coracle and the *Queen Mary*!

The Road to Wigan Pier, London: Gollancz (1937), 220–1, 235–7, 240, 241.

CHAPTER 12

Peaceful protest

G. Fox, *The journal of George Fox*

P. B. Shelley, *The Masque of Anarchy*

H. D. Thoreau, *Civil disobedience*

L. Tolstoy, *The abolition of government*

Introduction

> the decentralist counter-tendencies which can be perceived underlying all
> economic and social evolution [act in alliance with] something that is slowly
> evolving in the human soul: the most intimate of all resistances – resistance to
> mass or collective loneliness.

(Buber 1968: 14)

Greens and their predecessors have endorsed struggle against environmental and human
degradation, but have tended to believe that such a campaign should be non-violent.
Fundamental to their philosophy has been a revolt against authority and an emphasis on
creating face-to-face grass-roots democracy. The Green movement is keenly decentralist in
its vision of a better society. Non-violent direct action is practised as a form of opposition
to environmental destruction and militarism. In particular a major influence on the growth
of European Green politics has been the campaign launched in the early 1980s against the
deployment of cruise missiles. German Greens such as Petra Kelly were at the forefront of
this protest movement, which drew in millions of activists all over Europe. The women's
protest at Greenham Common provided particular inspiration, uniting radical left politics,
feminism and pagan spirituality. Several Greenham women contested the 1983 general
election as Women for Life on Earth/Ecology candidates. Greenham built upon a tradition
of non-violent women's protest, which included the Suffragettes' struggle for the vote in
the Edwardian era, as well as the efforts of women peace campaigners during the First
World War and in Cold War America. A Gandhian strategy transmitted via the Committee
of One Hundred, led by the philosopher Bertrand Russell, had earlier given rise to direct
action in the 1960s.

The Green movement, with its belief in grass-roots democracy, despite its parliamentary pretensions, has always been close to an anarchist tradition. Gandhi, Tolstoy and
Thoreau, men from three different continents, basing their politics on varied but deeply
held religious commitment – Hinduism, Christianity and Pantheism – were all anarchists
in the sense that they fought authority and its injustices with non-violent protest. All three
can be described as Greens, supporting animal rights, campaigning for social justice,
opposed to industrial capitalism and fervently anti-materialistic. Radical Christians have
argued that Christ was a radical in such a tradition and inspired the rest with his message
of 'turning the other cheek'. Here we see in our first extract a classic spiritual defence of
non-violence combined with a far from soft-spoken challenge to authority from George
Fox (1624–91), founder of the Quakers. Although it would be unrealistic to label Fox a

proto-Green, the importance of the Quakers as precursors and originators of the modern Green movement cannot be ignored. His Society of Friends, founded after he had suffered extreme spiritual unrest in 1646, drew in many former adherents of the anarchist and often mystically ecological sects common in the seventeenth century. As well as many Anabaptists, Familists and Ranters, the revolutionary pacifist Gerard Winstanley may have become a Quaker (Chapter Seventeen). The rejection of all elites, and belief in direct communion between the individual and God, gave the Quakers a strongly moral anti-authoritarian tendency.

Shelley, whom we have already referred to, was not only a vegetarian radical and ecologically sensitive critic of Malthus but a prominent pioneer of non-violence. Gandhi and Tolstoy claimed to have been inspired by his poem *The Masque of Anarchy.* 'Written on the occasion of the massacre at Manchester', it recounts how a 1819 working-class protest was violently put down by the army. The Peterloo massacre was the Amritsar or Sharpeville of its day. The monarch 'On a white horse splashed with blood' and his greedy servants Castlereagh, Eldon, Sidmouth, should be countered 'With folded arms and steady eyes, And little fear and less surprise'. Mass non-violence will work, according to the poet, because 'Ye are many – they are few.'

Next we find Thoreau (1817–62) accepting in defiant tones the statement 'That government is best which governs least', imprisoned for refusing to pay his 'poll tax' in protest at slavery. For Thoreau the payment of tax enabled the 'State to commit violence and shed innocent blood'. Non-payment defines 'a peaceable revolution, if any such is possible'. Tolstoy briefly reminds us that we shall be free only with the abolition of governments and nations, echoing the utopian wish of radical Greens.

George Fox

The journal of George Fox

Now the time of my commitment to the house of correction being nearly ended, and there being many new soldiers raised, the commissioners would have made me captain over them; and the soldiers said they would have none but me. So the keeper of the house of correction was commanded to bring me before the commissioners and soldiers in the market-place; and there they offered me that preferment, as they called it, asking me, if I would not take up arms for the Commonwealth against Charles Stuart? I told them, I knew from whence all wars arose, even from lust, according to James's doctrine; and that I lived in the virtue of that life and power that took away the occasion of all wars. But they courted me to accept their offer, and thought I did but compliment them. But I told them, I was come into the covenant of peace, which was before wars and strife were. They said, they offered it in love and kindness to me, because of my virtue; and such like flattering words they used. But I told them, if that was their love and kindness, I trampled it under my feet. Then their rage got up, and they said, 'Take him away, jailor, and put him into the dungeon amongst the rogues and felons.' So I was had away and put into a lousy, stinking place, without any bed, amongst thirty felons, where I was kept almost half a year, unless it were at times; for they would sometimes let me walk in the garden, having a belief that I would not go away. Now when they had got me into Derby dungeon it was the belief and saying of the people that I should never come out; but I had faith in God, and believed I should be delivered in his time; for the Lord had said to me before, that I was not to be removed from the place yet, being set there for a service which he had for me to do.

I was moved of the Lord to write a paper to the Protector, Oliver Cromwell; 'Wherein I did in the presence of the Lord God declare, that I denied the wearing or drawing of a carnal sword, or any other outward weapon, against him or any man: and that I was sent of God to stand a witness against all violence, and against the works of darkness; and to turn people from darkness to light; and to bring them from the causes of war and fighting, to the peaceable gospel, and from being evil-doers, which the magistrate's swords should be a terror to.'

All that pretend to fight for Christ, are deceived; for his kingdom is not of this world, therefore his servants do not fight. Fighters are not of Christ's kingdom, but are without Christ's kingdom; his kingdom starts in peace, and righteousness, but fighters are in the lust; and all that would destroy men's lives, are not of Christ's mind, who came to save men's lives. Christ's kingdom is not of this world; it is peaceable: and all that are in strife, are not of his kingdom. All that pretend to fight for the Gospel, are deceived; for the gospel is the power of God, which was before the devil, or fall of man was; and the gospel of peace was before fighting was. Therefore they that pretend fighting, are ignorant of the gospel; and all that talk of fighting for Sion, are in darkness; for Sion needs no such helpers. All such as profess themselves to be ministers of Christ, or Christians, and go about to beat down the whore with outward, carnal weapons, the flesh and the whore are got up in themselves, and they are in a blind zeal; for the whore is got up by the inward ravening from the Spirit of God; and the beating down thereof, must be by the inward stroke of the sword of the Spirit within. All such as pretend Jesus Christ, and confess him, and yet run into the use of carnal weapons, wrestling with flesh and blood, throw away the spiritual weapons. They that would be wrestlers with flesh and blood, throw away Christ's doctrine; the flesh is got up in them, and they are weary of their sufferings. Such as would revenge themselves, are out of Christ's doctrine. Such as being stricken on one cheek, would not turn the other, are out of Christ's doctrine: and such as do not love one another, nor love enemies, are out of Christ's doctrine.

*Christians are commanded to love enemies; therefore much more, one another. And Christ saith, 'As the Father hath loved me, so I have loved you: continue ye in my love,' John xv. 8, and 'By this shall all men know that ye are my disciples, if ye love one another,' John xiii. 35. But if Christians war and destroy one another, this will make Jews, Turks, Tartars, and Heathens say, that you are not disciples of Christ. Therefore, as you love God, and Christ, and Christianity, and its peace, all make peace, as far as you have power, among Christians, that you may have blessing. You read, that Christians were called the household of faith, the household of God, a holy nation, a peculiar people: and they are commanded to be 'zealous of good works', not of bad: and Christians are also commanded not to bite and devour one another, lest they be consumed one of another.

Peaceful protest

It is not a sad thing for Christians to be biting and consuming one another in the sight of the Turks, Tartars, Jews, and Heathens, when they should 'love one another, and do unto all men, as they would have them do unto them?' Such devouring work as this will open the mouths of Jews and Turks, Tartars and Heathens, to blaspheme the name of Christ, who is King of kings, and Lord of lords, and cause them to speak evil of Christianity, for them to see how the unity of the Spirit is broken among such, as profess Christ and Christ's peace.

Note

*[From an Epistle to 'the Ambassadors, that are met to treat for Peace at the City of Nimeguan, in the States' dominions, July 21, 1677'.]

The Journal of George Fox (1694), London: Friends' Tract Association (1901), 1: 68, 69, 209, 448.

Percy Bysshe Shelley

The Masque of Anarchy

I

As I lay asleep in Italy,
There came a voice from over the sea,
And with great power it forth led me
To walk in the visions of Poesy.

II

I met Murder on the way –
He had a mask like Castlereagh.
Very smooth he looked, yet grim;
Seven bloodhounds followed him.

III

All were fat; and well they might
Be in admirable plight,
For one by one, and two by two,
He tossed them human hearts to chew,
Which from his wide cloak he drew.

Green history

IV

Next came Fraud, and he had on,
Like Eldon, an ermined gown.
His big tears, for he wept well,
Turned to millstones as they fell;

V

And the little children who
Round his feet played to and fro,
Thinking every tear a gem,
Had their brains knocked out by them.

VI

Clothed with the bible, as with light
And the shadows of the night,
Like Sidmouth next, Hypocrisy
On a crocodile rode by.

VII

And many more Destructions played
In this ghastly masquerade, –
All disguised, even to the eyes,
Like bishops, lawyers, peers, or spies.

VIII

Last came Anarchy; he rode
On a white horse splashed with blood;
He was pale even to the lips,
Like Death in the Apocalypse.

IX

And he wore a kingly crown,
And in his grasp a sceptre shone;
On his brow this mark I saw –
'I AM GOD AND KING, AND LAW.'

X

With a pace stately and fast
Over English land he passed,
Trampling to a mire of blood
The adoring multitude.

Peaceful protest

XI

And a might troop around
With their trampling shook the ground,
Waving each a bloody sword
For the service of their lord.

XII

And with glorious triumph they
Rode through England, proud and gay,
Drunk as with intoxication
Of the wine of desolation.

XIII

O'er fields and towns, from sea to sea,
Passed the pageant swift and free,
Tearing up and trampling down,
Till they came to London town.

XIV

And each dweller, panic-stricken,
Felt his heart with terror sicken,
Hearing the tempestuous cry
Of the triumph of Anarchy.

XV

For with pomp to meet him came,
Clothed in arms like blood and flame,
The hired murderers who did sing,
'Thou art God, and Law, and King!'

XVI

'We have waited, weak and lone,
For thy coming, Mighty One!
Our purses are empty, our swords are cold;
Give us glory, and blood, and gold.'

XVII

Lawyers and priests, a motley crowd,
To the earth their pale brows bowed –
Like a bad prayer not over-loud,
Whispering 'Thou art Law and God!'

Green history

XVIII

Then all cried with one accord,
'Thou art King, and God, and Lord;
Anarchy, to thee we bow;
Be thy name made holy now!'

XIX

And Anarchy the skeleton
Bowed and grinned to every one
As well as if his education
Had cost ten millions to the nation.

XX

For he knew the palaces
Of our kings were nightly his;
His the sceptre, crown, and globe,
And the gold-inwoven robe. . . .

LXIII

'Or turn their wealth to arms, and make
War, for thy beloved sake,
On wealth and war and fraud, whence they
Drew the power which is their prey.

LXIV

'Science, poetry, and thought,
Are thy lamps; they make the lot
Of the dwellers in a cot
Such they curse their Maker not.

LXV

'Spirit, patience, gentleness,
All that can adorn and bless,
Art thou. Let deeds, not words, express
Thine exceeding loveliness.

LXVI

'Let a great assembly be
Of the fearless and the free
On some spot of English ground
Where the plains stretch wide around.

Peaceful protest

LXVII

'Let the blue sky overhead,
The green earth on which ye tread,
All that must eternal be,
Witness the solemnity.

LXVIII

'From the corners uttermost
Of the bounds of English coast;
From every hut, village, and town,
Where those who live and suffer moan
For others' misery or their own;

LXIX

'From the workhouse and the prison
Where, pale as corpses newly risen,
Women, children, young and old,
Groan for pain, and weep for cold;

LXX

'From the haunts of daily life
Where is waged the daily strife
With common wants and common cares
Which sows the human heart with tares;

LXXI

'Lastly, from the palaces
Where the murmur of distress
Echoes like the distant sound
Of a wind alive around

LXXII

'Those prison-halls of wealth and fashion
Where some few feel such compassion,
For those who groan and toil and wail,
As must make their brethren pale; –

LXXIII

'Ye who suffer woes untold
Or to feel or to behold
Your lost country bought and sold
With a price of blood and gold; –

LXXIV

'Let a vast assembly be,
And with great solemnity
Declare with measured words that ye
Are, as God has made ye, free!

LXXV

'Be your strong and simple words
Keen to wound as sharpened swords,
And wide as targes let them be,
With their shade to cover ye.

LXXVI

'Let the tyrants pour around
With a quick and startling sound,
Like the loosening of a sea,
Troops of armed emblazonry.

LXXVII

'Let the charged artillery drive,
Till the dead air seems alive
With the clash of clanging wheels,
And the tramp of horses' heels.

LXXVIII

'Let the fixèd bayonet
Gleam with sharp desire to wet
Its bright point in English blood,
Looking keen as one for food.

LXXIX

'Let the horsemen's scimitars
Wheel and flash, like sphereless stars
Thirsting to eclipse their burning
In a sea of death and mourning.

LXXX

'Stand ye calm and resolute,
Like a forest close and mute,
With folded arms, and looks which are
Weapons of an unvanquished war.

Peaceful protest

LXXXI

'And let Panic, who outspeeds
The career of armèd steeds,
Pass, a disregarded shade,
Through your phalanx undismayed.

LXXXII

'Let the laws of your own land,
Good or ill, between ye stand,
Hand to hand, and foot to foot,
Arbiters of the dispute: –

LXXXIII

'The old laws of England – they
Whose reverend heads with age are grey,
Children of a wiser day;
And whose solemn voice must be
Thine own echo – Liberty!

LXXXIV

'On those who first should violate
Such sacred heralds in their state
Rest the blood that must ensue;
And it will not rest on you.

LXXXV

'And, if then the tyrants dare,
Let them ride among you there,
Slash and stab and maim and hew:
What they like, that let them do.

LXXXVI

'With folded arms and steady eyes,
And little fear and less surprise,
Look upon them as they slay,
Till their rage has died away.

LXXXVII

'Then they will return with shame
To the place from which they came,
And the blood thus shed will speak
In hot blushes on their cheek.

LXXXVIII

'Every woman in the land
Will point at them as they stand –
They will hardly dare to greet
Their acquaintance in the street:

LXXXIX

'And the bold true warriors
Who have hugged danger in wars
Will turn to those who would be free,
Ashamed of such base company:

XC

'And that slaughter to the nation
Shall steam up like inspiration,
Eloquent, oracular,
A volcano heard afar;

XCI

'And these words shall then become
Like Oppression's thundered doom,
Ringing through each heart and brain,
Heard again – again – again!

XCII

'Rise, like lions after slumber,
In unvanquishable number!
Shake your chains to earth, like dew
Which in sleep had fallen on you!
Ye are many – they are few.'

The Masque of Anarchy (1819), *The Complete Poetical Works of Percy Bysshe Shelley*, London: Gibbings (1894), 261, 262, 270–2.

Henry David Thoreau

Civil disobedience

I heartily accept the motto, 'That government is best which governs least,' and I should like to see it acted up to more rapidly and systematically. Carried out, it finally amounts to this, which also I believe – 'That government is best which governs not at all'; and when men are prepared for it, that will be the kind of government which they will have. Government is at best but an expedient; but most governments are usually, and all governments are sometimes, inexpedient. The objections which have been brought against a standing army, and they are many and weighty, and deserve to prevail, may also at last be brought against a standing government. The standing army is only an arm of the standing government. The government itself, which is only the mode which the people have chosen to execute their will, is equally liable to be abused and perverted before the people can act through it. Witness the present Mexican war, the work of comparatively a few individuals using the standing government as their tool; for, in the outset, the people would not have consented to this measure.

This American government – what is it but a tradition, though a recent one, endeavoring to transmit itself unimpaired to posterity, but each instant losing some of its integrity? It has not the vitality and force of a single living man, for a single man can bend it to his will. It is a sort of wooden gun to the people themselves. But it is not the less necessary for this, for the people must have some complicated machinery or other, and hear its din, to satsify the idea of government which they have. Governments show thus how successfully men can be imposed on, even impose on themselves, for their own advantage. It is excellent, we must all allow. Yet this government never of itself furthered any enterprise, but by the alacrity with which it got out of its way. *It* does not keep the country free. *It* does not settle the West. *It* does not educate. The character inherent in the American people has done all that has been accomplished, and it would have done somewhat more if the government had not sometimes got in its way. . . .

How does it become a man to behave toward this American government today? I answer, that he cannot without disgrace be associated with it. I cannot for an instant recognize that political organization as *my* government which is the *slave*'s government also.

All men recognize the right of revolution; that is, the right to refuse allegiance to, and to resist, the government, when its tyranny or its inefficiency are great and unendurable. But almost all say that such is not the case now. But such was the case, they think, in the Revolution of '75. If one were to tell me that this was a bad government because it taxed certain foreign commodities brought to its ports, it is most probable that I should not make an ado about it, for I can do without them. All machines have their friction, and possibly this does enough good to counterbalance the evil. At any rate, it is a great evil to make a stir about it. But when the friction comes to have its machine, and oppression and robbery are organized, I say, let us not have such a machine any longer. In other words, when a sixth of the population of a nation which has undertaken to be the refuge of liberty are slaves, and a whole country is unjustly overrun and conquered by a foreign army and subjected to military law, I think that it is not too soon for honest men to rebel and revolutionize. What makes this duty the more urgent is the fact that the country so overrun is not our own, but ours is the invading army. . . .

. . . If my esteemed neighbor, the State's ambassador, who will devote his days to the settlement of the question of human rights in the Council Chamber instead of being threatened with the prisons of Carolina, were to sit down the prisoner of Massachusetts, that State which is so anxious to foist the sin of slavery upon her sister – though at present she can discover only an act of inhospitality to be the ground of a quarrel with her

– the Legislature would not wholly waive the subject the following winter.

Under a government which imprisons any unjustly, the true place for a just man is also a prison. The proper place today, the only place which Massachusetts has provided for her freer and less despondent spirits, is in her prisons, to be put out and locked out of the State by her own act, as they have already put themselves out by their principles. It is there that the fugitive slave, and the Mexican prisoner on parole, and the Indian come to plead the wrongs of his race should find them; on that separate, but more free and honorable ground where the State places those who are not *with* her, but *against* her – the only house in a slave State in which a free man can abide with honor. If any think that their influence would be lost there, and their voices no longer afflict the ear of the State, that they would not be as an enemy within its walls, they do not know by how much truth is stronger than error, nor how much more eloquently and effectively he can combat injustice who has experienced a little in his own person. Cast your whole vote, not a strip of paper merely, but your whole influence. A minority is powerless while it conforms to the majority; it is not even a minority then; but it is irresistible when it clogs by its whole weight. If the alternative is to keep all just men in prison, or give up war and slavery, the State will not hesitate which to choose. If a thousand men were not to pay their tax-bills this year, that would not be a violent and bloody measure as it would be to pay them and enable the State to commit violence and shed innocent blood. This is, in fact, the definition of a peaceable revolution, if any such is possible.

Civil Disobedience (1849), Westwood, N.J.: Revell (1964), 11–12, 17–18, 31–3.

Leo Tolstoy

The abolition of government

The abolition of government will merely rid us of an unnecessary organisation which we have inherited from the past for the commission of violence and for its justification. 'But there will be no laws, no property, no courts of justice, no police, no popular education,' say people who intentionally confuse the use of violence by governments with various social activities. The abolition of the organisation of government formed to do violence does not at all involve the abolition of what is reasonable and good, and therefore not based on violence, in laws or law courts, or in property, or in popular education. On the contrary, the absence of the brutal power of government, which is needed only for its own supports, will facilitate a more just and reasonable social organisation, needing no violence. Courts of justice and public affairs, and popular education, will all exist to the extent to which they are really needed by the people, but in a shape which will not involve the evils contained in the present form of government. What will be destroyed is merely what was evil and hindered the free expression of the people's will.

Understand that salvation from your woes is only possible when you free yourself from the obsolete idea of patriotism and from obedience to governments that is based upon it, and when you boldly enter into the region of that higher ideal, the brotherly union of the peoples, which has long since come to life and from all sides is calling yourself to itself.

The Life and Teaching of Leo Tolstoy, London: Grant Richard (1901), 158–9.

CHAPTER 13

The city and the country

W. Cobbett, Cottage economy

E. Howard, Tomorrow

P. Kropotkin, Fields, factories and workshops

L. Mumford, The culture of cities

A. Walker, Longing to die of old age

Introduction

> Fritz was so impressed by the Soil Association that he at once began to imple-
> ment organic methods in his own back garden. He made enormous compost
> heaps, offending the neighbours by the cartloads of manure he imported from
> a nearby pig farm to improve the quality of the chalky soil.
>
> (Wood 1984: 221–2)

The Green movement has been defined as ruralist, concerned not just with environmental quality but with promoting a way of life close to the land. Peasant revolutionaries from John Ball (d. 1381) in the thirteenth century to the Chartist Fergus O'Connor (1794–1855) have called for radical land reform. In Eastern Europe peasant parties, organized in a 'Green International', with similar demands were prominent in the 1920s, and land reform was one of Gandhi's principal concerns while campaigning to remove the British from India. The movement for allotments and the current practice of situating city farms in depressed urban areas are relatively modern remnants of a vigorous 'back to the land' campaign.

Organic cultivation, opposition to the use of green-field sites for building and soil conservation measures have all been fought for by the Green movement. 'Utopian' socialists like Owen argued that productivity could be increased while maintaining soil fertility if more people were put to work on the land. Rural depopulation and animal rights issues, especially those linked with factory farming, are more recent concerns. William Cobbett (1763–1835) hated 'the monstrous Wen', as he described London, and advised the smallholder in his *Cottage Economy*. With his detailed prescriptions for bread making, pig rearing and brewing, he attempts to provide the practical details of how the sturdy independent peasant may maintain his status.

There has always been a dialectic between attempts to maintain or recreate rural balance with attempts to reform the cities. The present debate between Trainer, who sees the need to 'green' cities, and supporters of anti-urbanism such as Earth First! is paralleled by centuries of similar debate (Trainer 1985). The bitterly anti-urban Narodniks felt that the growth of cities had led directly to the impoverishment of rural Russia. Many Romantic poets attacked the cities as innately against nature and did not just praise the cultivated countryside but looked for their inspiration to supposedly untouched wilderness. Trainer argues that cities can and must be made 'green', believing that the abolition of the car and the creation of vast city farms and parks will provide the solution to both social and ecological problems. Engels argued that cities had no long-term future while they failed to give their manure back to the land; although by no means an obvious friend of the peasant, Engels states, without giving any detailed prescription, the need to combine both city and country. We hear from the anarchist prince and friend of William Morris, Peter Kropotkin (1842–1921). Providing the classic account of a working Green economy, his system combines agriculture and light industry, providing a bridge between perhaps William Morris's *News from Nowhere* and everyday reality. Kropotkin's emphasis on decentralization also clearly inspired *Blueprint for Survival* vision of an ecological sustainable twenty-first century Britain.

Ebenezer Howard (1850–1928), the first in a line of radical town planners that culminates with Mumford, proposes in exact detail how such a seemingly 'utopian' project might be made to work. He optimistically proclaims that 'Town and country must be married, and out of this joyous union will spring a new hope, a new life, a new civilisation'. Although in Britain a modest Garden Cities movement sprang from his writings, his views were ignored. Welwyn Garden City and Hampstead Garden Suburb may have been leafy but they lacked an agricultural component and peasant workers. In the Soviet Union, under Lenin and Stalin, the anti-environmental and anti-peasant strands in Marx's and Engels's writings came to the fore. Mumford, who died in 1990 but began writing in the 1920s, bridges the gap between Howard and the contemporary Greens. Here in 1938 he attacks the pollution and waste of the vast urban 'agglomeration', in particular examining the pernicious influence of the car.

Finally, the American novelist Alice Walker, author of *The Color Purple*, describes how her 4-greats' grandmother lived to the grand age of 121 on a diet of home produce. Mrs Mary Poole, formerly a slave, born in 1800, died in 1921, her longevity ascribed to a robust, self-sufficient peasant lifestyle. The description of her life illustrates, without romanticism, a process foreseen by Cobbett and experienced by African Americans, Ukrainian Jews, English and agricultural smallholders displaced from the country to bleaker forms of urban life.

William Cobbett

Cottage economy

2. The word Economy, like a great many others, has, in its application, been very much abused. It is generally used as if it meant parsimony, stinginess, or niggardliness; and, at best, merely the refraining from expending money. Hence misers and close-fisted men disguise their propensity and conduct under the name of economy; whereas the most liberal disposition, a disposition precisely the contrary of that of the miser, is perfectly consistent with economy.

3. ECONOMY means management, and nothing more; and it is generally applied to the affairs of a house and family, which affairs are an object of the greatest importance, whether as relating to individuals or to a nation. A nation is made powerful and to be honoured in the world, not so much by the number of its people as by the ability and

character of that people; and the ability and character of a people depend, in a great measure, upon the economy of the several families, which, all taken together, make up the nation. There never yet was, and never will be, a nation permanently great, consisting, for the greater part, of wretched and miserable families.

4. In every view of the matter, therefore, it is desirable, that the families of which a nation consists should be happily off; and as this depends, in a great degree, upon the management of their concerns, the present work is intended to convey, to the families of the labouring classes in particular, such information as I think may be useful with regard to that management.

5. I lay it down as a maxim, that for a family to be happy, they must be well supplied with food and raiment. It is a sorry effort that people make to persuade others, or to persuade themselves, that they can be happy in a state of want of the necessaries of life. The doctrines which fanaticism preaches, and which teach men to be content with poverty, have a very pernicious tendency, and are calculated to favour tyrants by giving them passive slaves. To live well, to enjoy all things that make life pleasant, is the right of every man who constantly uses his strength judiciously and lawfully. It is to blaspheme God to suppose that he created men to be miserable, to hunger, thirst, and perish with cold, in the midst of that abundance which is the fruit of their own labour. Instead, therefore, of applauding 'happy poverty', which applause is so much the fashion of the present day, I despise the man that is poor and contented; for such content is a certain proof of a base disposition, a disposition which is the enemy of all industry, all exertion, all love of independence.

6. Let it be understood, however, that, by poverty, I mean real want, a real in-sufficiency of the food and raiment and lodging necessary to health and decency; and not that imaginary poverty, of which some persons complain. The man who, by his own and his family's labour, can provide a sufficiency of food and raiment, and a comfortable dwelling place, is not a poor man. There must be different ranks and degrees in every civil society, and, indeed, so it is even amongst the savage tribes. There must be different degrees of wealth; some must have more than others; and the richest must be a great deal richer than the least rich. But it is necessary to the very existence of a people, that nine out of ten should live wholly by the sweat of their brow; and, is it not degrading to human nature, that all the nine-tenths should be called poor; and, what is still worse, call themselves poor, and be contented in that degraded state?

7. The laws, the economy, or management, of a state may be such as to render it impossible for the labourer, however skilful and industrious, to maintain his family in health and decency; and such has, for many years past, been the management of the affairs of this once truly great and happy land. A system of paper money, the effect of which was to take from the labourer the half of his earnings, was what no industry and care could make head against. I do not pretend that this system was adopted by design. But, no matter for the cause; such was the effect.

8. Better times, however, are approaching. The labourer now appears likely to obtain that hire of which he is worthy; and, therefore, this appears to me to be the time to press upon him the duty of using his best exertions for the rearing of his family in a manner that must give him the best security for happiness to himself, his wife and children, and to make him, in all respects, what his forefathers were. The people of England have been famed, in all ages, for their good living; for the abundance of their food, and goodness of

their attire. The old sayings about English roast beef and plum-pudding, and about English hospitality, had not their foundation in *nothing*. And in spite of all refinements of sickly minds, it is *abundant living* amongst the people at large, which is the great test of good government, and the surest basis of national greatness and security. . . .

16. But, the basis of good to him, is steady and skilful labour. To assist him in the pursuit of this labour, and in the turning of it to the best account, are the principal objects of the present little work. I propose to treat of brewing Beer, making Bread, keeping Cows and Pigs, rearing Poultry, and of other matters; and to show that, while, from a very small piece of ground, a large part of the food of a considerable family may be raised, the very act of raising it will be the best possible foundation of the education of the children of the labourer; that it will teach them a great number of useful things, add greatly to their value when they go forth from their father's home, make them start in life with all possible advantages, and give them the best chance of leading happy lives. And is it not much more rational for parents to be employed in teaching their children how to cultivate a garden, to feed and rear animals, to make bread, beer, bacon, butter, and cheese, and to be able to do these things for themselves, or for others, than to leave them to prowl about the lanes and commons, or to mope at the heels of some crafty, sleek-headed pretended saint, who while he extracts the last penny from their pockets, bids them be contented with their misery, and promises them, in exchange for their pence, everlasting glory in the world to come? It is upon the hungry and wretched that the fanatic works. The dejected and forlorn are his prey. As an ailing carcass engenders vermin, a pauperised community engenders teachers of fanaticism, the very foundation of whose doctrines is, that we are to care nothing about this world, and that all our labourers and exertions are in vain.

Cottage Economy (1821), London: Peter Davies (1926), 1–4, 8–9.

Ebenezer Howard

Tomorrow

There are in reality not only, as is so constantly assumed, two alternatives – town life and country life – but a third alternative, in which all the advantages of the most energetic and active town life, with all the beauty and delight of the country, may be secured in perfect combination; and the certainty of being able to live this life will be the magnet which will produce the effect for which we are all striving – the spontaneous movement of the people from our crowded cities to the bosom of our kindly mother earth, at once the source of life, of happiness, of wealth, and of power. The town and the country may, therefore, be regarded as two magnets, each striving to draw the people to itself – a rivalry which a new form of life, partaking of the nature of both, comes to take part in. This may be illustrated by a diagram of 'The Three Magnets', in which the chief advantages of the Town and of the Country are set forth with their corresponding drawbacks, while the advantages of the Town–Country are seen to be free from the disadvantages of either. . . .

But neither the Town magnet nor the Country magnet represents the full plan and purpose of nature. Human society and the beauty of nature are meant to be enjoyed together. The two magnets must be made one. As man and woman by their varied gifts and faculties supplement each other, so should town and country. The town is the symbol of society – of mutual help and friendly co-operation, of fatherhood, motherhood, brother-hood, sisterhood, of wide relations between man and man – of broad, expanding sympathies – of science, art, culture, religion. And the country! The country is the symbol of God's love and care for man. All that we are, and all that we have comes from it. Our bodies are formed of it; to it they return.

Tomorrow, London: Swan Sonnenschein (1898), 7, 9.

Peter Kropotkin

Fields, factories and workshops

In the domain of agriculture it may be taken as proved that if a small part only of the time that is now given in each nation or region to field culture was given to well thought out and socially carried out permanent improvements of the soil, the duration of work which would be required afterwards to grow the yearly bread-food for an average family of five would be less than a fortnight every year; and that the work required for that purpose would not be the hard toil of the ancient slave, but work which would be agreeable to the physical forces of every healthy man and woman in the country.

It has been proved that by the following methods of intensive market-gardening – partly under glass – vegetables and fruit can be grown in such quantities that men could be provided with rich food and a profusion of fruit, if they simply devoted to the task of growing them the hours which everyone willingly devotes to work in the open air, after having spent most of his day in the factory, the mine or the study. Provided, of course, that the production of foodstuffs should not be the work of the isolated individual, but the planned-out and combined action of human groups.

It has also been proved – and those who care to verify it by themselves may easily do so by calculating the real expenditure for labour which was lately made in the building of workmen's houses by both private persons and municipalities – that under a proper combination of labour, twenty to twenty-four months of one man's work would be sufficient to secure for ever, for a family of five, an apartment or a house provided with all the comforts which modern hygiene and taste could require.

And it has been demonstrated by actual experiment that, by adopting methods of education, advocated long since and partially applied here and there, it is most easy to convey to children of an average intelligence, before they have reached the age of fourteen or fifteen, a broad general comprehension of Nature, as well as of human societies; to familiarise their minds with sound methods of both scientific research and technical work, and to inspire their hearts with a deep feeling of human solidarity and justice; and that is extremely easy to convey during the next four or five years a reasoned scientific knowledge of Nature's laws, as well as a knowledge, at once reasoned and practical, of the technical methods of satisfying man's material needs. Far from being inferior to the 'specialised' young persons manufactured by our universities, the *complete* human being, trained to use his brain and his hands, excels them, on the contrary, in all respects, especially as an initiator and an inventor in both science and technics.

All this has been proved. It is an acquisition of the times we live in – an acquisition which has been won despite the innumerable obstacles always thrown in the way of every initiative mind. It has been won by the obscure tillers of the soil, from whose hands greedy states, landlords and middlemen snatch the fruits of their labour even before it is ripe; by obscure teachers who only too often fall crushed under the weight of church, state, commercial competition, inertia of mind and prejudice.

And now, in the presence of all this conquest – what is the reality of things?

Nine-tenths of the whole population of grain-exporting countries like Russia, one-half of it in countries like France which live on home-grown food, work upon the land – most of them in the same way as the slaves of antiquity did, only to obtain a meagre crop from a soil, and with a machinery which they cannot improve, because taxation, rent and usury keep them always as near as possible to the margin of starvation. At the beginning of this century, whole populations plough with the same plough as their medieval ancestors, live in the same incertitude of the morrow, and are as carefully denied eduction; and they have, in claiming their portion of bread, to march with their children and wives against their own sons' bayonets, as their grandfathers did hundreds of years ago.

In industrially developed countries, a couple of months' work, or even much less than that, would be sufficient to produce for a family a rich and varied vegetable and animal food. But the researchers of Engel [at Berlin] and his many followers tell us that the workman's family has to spend one full half of its yearly earnings – that is, to give six months of labour, and often more – to provide its food. And what food! Is not bread and dripping the staple food of more than one-half of English children?

One month of work every year would be quite sufficient to provide the worker with a healthy dwelling. But it is from 25 to 40 per cent of his yearly earnings – that is, from three to five months of his working time every year – that he has to spend in order to get a dwelling, in most cases unhealthy and far too small; and this dwelling will never be his own, even though at the age of forty-five or fifty he is sure to be sent away from the factory, because the work that he used to do will by that time be accomplished by a machine and a child.

We all know that the child ought, at least, to be familiarised with the forces of Nature which some day he will have to utilise, ... Everyone will grant thus much; but what do we do? From the age of ten or even nine we send the child to push a coal-cart in a mine, or to bind, with a little monkey's agility, the two ends of threads broken in a spinning gin. From the age of thirteen we compel the girl – a child yet – to work as a 'woman' at the weaving-loom, or to stew in the poisoned, overheated air of a cotton-dressing factory, or, perhaps, to be poisoned in the death chambers of a Staffordshire pottery.... What floods of useless sufferings deluge every so-called civilised land in the world!

When we look back on ages past, and see there the same sufferings, we may say that perhaps then they were unavoidable on account of the ignorance which prevailed. But

171

human genius, stimulated by our modern Renaissance, had already indicated new paths to follow.

For thousands of years in succession, to grow one's own food was the burden, almost the curse, of mankind. But it need be so no more. If you make yourselves the soil, and partly the temperature and the moisture which each crop requires, you will see that to grow the yearly food of a family, under rational conditions of culture, requires so little labour that it might almost be done as a mere change from other pursuits. If you return to the soil, and co-operate with your neighbours instead of erecting high walls to conceal yourself from their looks; if you utilise what experiment has already taught us ... you will be astonished at the facility with which you can bring a rich and varied food out of the soil.

... Have the factory and the workshop at the gates of your fields and gardens, and work in them. Not those large establishments, of course, in which huge masses of metals have to be dealt with and which are better placed at certain spots indicated by Nature, but the countless variety of workshops and factories which are required to satisfy the infinite diversity of tastes among civilised men. Not those factories in which children lose all the appearance of children in the atmosphere of an industrial hell, but those airy and hygienic, and consequently economical, factories in which human life is of more account than machinery and of extra profits.... Let those factories be erected, not for making profits by selling shoddy or useless and noxious things to the enslaved Africans, but to satisfy the unsatisfied needs of millions of Europeans.... Very soon you will yourselves feel interested in that work, and you will have occasion to admire in your children their eager desire to become acquainted with Nature and its forces, their inquisitive inquiries as the powers of machinery, and their rapidly developing inventive genius....

Fields, Factories and Workshops, London: Swann Sonnenschein (1901), 213–18.

Lewis Mumford

The culture of cities

Meanwhile, the urban agglomeration produces a similar depletion in the natural environment. Nature, except in a surviving landscape park, is scarcely to be found near the metropolis: if at all, one must look overhead, at the clouds, the sun, the moon, when they appear through the jutting towers and building blocks. The blare of light in the evening sky blots out half the stars overhead: the rush of sewage into the surrounding waters converts rivers into open sewers, drives away the more delicate feeders among the fish, and infects the bathers in the waters with typhoid: through the greater part of the nineteenth century typhoid was an endemic disease in big cities, brought in with the food supply, the shellfish, if not absorbed directly from the colon bacilli in the bathing or drinking water.

The city and the country

If the metropolis attempts to counteract these evils, it can do so only at a vast outlay: stations where the water is filtered and chlorinated, plants where the sewage is reduced and converted into fertilizer bring additional items of expense to the budget. If some isolated beauty in nature is preserved as a park, like the Bear Mountain Park, outside New York, it will be at a distance that requires half a morning to reach, even from the center of the city. When one arrives there one will find that a multitude of other people, equally eager to escape the metropolis, have by their presence created another metropolis – if not a wilderness–slum. One will see nature through the interstices of their bodies.

Indeed, the only successful metropolitan recreation grounds are those that accept the fact of overcrowding and give it appropriate form: a Wannseebad in Berlin, or a Jones Beach on Long Island: a vast stretch of waterfront domed by a vaster sky, well organized, efficiently policed, with thousands of automobiles drawn up in ranks, giant pavilions, scores of assiduous life-guards on spidery towers, thousands of bathers basking in the sun and watching each other. A great mass spectacle: perhaps the nearest approach to genuine life, life esthetically intensified and ordered, that the metropolis offers.

As the pavement spreads, nature is pushed farther away: the whole routine divorces itself more completely from the soil, from the visible presence of life and growth and decay, birth and death: the slaughterhouse and the cemetery are equally remote, and their processes are equally hidden. The ecstatic greeting of life, the tragic celebration of death, linger on merely as mumbled forms in the surviving churches. The rhythm of the season disappears, or rather, it is no longer associated with natural events, except in print. Millions of people grow up in this metropolitan milieu who know no other environment than the city streets: people to whom the magic of life is represented, not by the miracles of birth and growth, but by placing a coin in a slot and drawing out a piece of candy or a prize. This divorce from nature has serious physiological dangers that the utmost scruples of medical care scarcely rectify. For all its boasted medical research, for all its real triumphs in lessening the incidence of disease and prolonging life, the city must bow to the country-side in the essentials of health: almost universally the expectation of life is greater in the latter, and the effect of deteriorative diseases is less.

But how find the country? The depletion of the metropolis does not stop at the legal boundaries of the metropolis: urban blight leads to rural blight. Since 1910 or thereabouts, the highways of motor traffic have begun to spread out from every metropolis in ever thickening and multiplying streams: these highways carry with them the environment of the metropolis: the paved highway, the filling station, the roadside slum, the ribbon development of houses, the roadhouse and cabaret. The farther and faster one travels, the more the life that accompanies one remains like that one has left behind. The same standardization of ugliness: the same mechanical substitutes: the same cockney indifference to nature: the same flippant attitude: the same celluloid pleasures and canned noise. A row of bungalows in the open country alongside an express highway is a metropolitan fact: so are the little heaps of week-end cabins by lake or stream or oceanside. Their density and concentration may not be greater than that of a rural village: but in their mode of life, their amusements, their frame of social reference, they are entirely metropolitan: hardly better or worse for being fifty miles away from the center.

Under this regime, every environment bears the same taint: its abiding picture of life is colored by the same newspapers, the same magazines, the same moving pictures, the same radio. Dependent upon the metropolitan markets for current cash, the outlying

farming regions, mining centers, and industrial areas are all under the sway of metropolitan interests. What is not metropolitan is either the original bequest of nature, often neglected, misused, rundown, or a relic of an historical past when the community once showed an autonomous and autochthonous life. But the rural regions and the provincial towns taste only the metropolitan skimmed milk: the cream has been mechanically separated for the benefit of the big city. The provincial town now faces a poverty, or at least an impecuniousness, that is without the vicarious enjoyments of the metropolis, and without the residue of philanthropies, trusts, foundations, which provide the hospitals and libraries and institutions of learning in the big city: residual pledges of a better life.

The inhabitants of these rural areas, indeed, are taught to despise their local history, to avoid their local language and their regional accents, in favor of the colorless language of metropolitan journalism: their local cooking reflects the gastronomic subterfuges of the suburban women's magazines; their songs and dances, if they survive, are elbowed off the dance floor; at best are given an audition at a metropolitan cabaret or radio station, where they are driven to an early death by universal repetition. The whole moral of this metropolitan regime is that one does not live, truly live, unless one lives in the metropolis or copies closely, abjectly, its ways. Expensive ways: ways that may be turned into monetary profits for the benefit of those who have a capital stake in the regime and who live in the light of its reflected glory. This moral is implanted by education, driven home by advertisement, spread by propaganda: *life means metropolitan life*. Not merely is the exodus to the city hastened, but the domination of the surviving countryside is assured: the same hand, as it were, writes the songs and lays down the terms for the mortgage.

In short: to scorn one's roots in one's own region and admiringly to pluck the paper flowers manufactured and sold by the metropolis becomes the whole duty of man. Though the physical radius of the metropolis may be only twenty or thirty miles, its effective radius is much greater: its blight is carried in the air, like the spores of a mold. The outcome is a world whose immense potential variety, first fully disclosed to man during the nineteenth century, has been sacrificed to a low metropolitan standardization. A rootless world, removed from the sources of life: a Plutonian world, in which living forms become frozen into metal: cities expanding to no purpose, cutting off the very trunk of their regional existence, defiling their own nest, reaching into the sky after the moon: more paper profits, more vicarious substitutes for life. Under this regime more and more power gets into hands of fewer and fewer people, ever further and further away from reality.

The Culture of Cities, London: Secker & Warburg (1938), 252–5.

Alice Walker

Longing to die of old age

Mrs. Mary Poole, my '4-greats' grandmother, lived the entire nineteenth century, from around 1800 to 1921, and enjoyed exceptional health. The key to good health, she taught (this woman who as an enslaved person was forced to carry two young children, on foot, from Virginia to Georgia), was never to cover up the pulse at the throat. But, with the benefit of hindsight, one must believe that for her, as for generations of people after her, in our small farming community, diet played as large a role in her longevity and her health as loose clothing and fresh air.

For what did the old ones eat?

Well, first of all, almost nothing that came from a store. As late as my own childhood, in the fifties, at Christmas we had only raisins and perhaps bananas, oranges, and a peppermint stick, broken into many pieces, a sliver for each child; and during the year, perhaps, a half-dozen apples, nuts, and a bunch of grapes. All extravagantly expensive and considered rare. You ate *all* of the apple, sometimes, even, the seeds. Everyone had a vegetable garden; a garden as large as there was energy to work it. In these gardens people raised an abundance of food: corn, tomatoes, okra, peas and beans, squash, peppers, which they ate in summer and canned for winter. There was no chemical fertilizer. No one could have afforded it, had it existed, and there was no need for it. From the cows and pigs and goats, horses, mules and fowl that people also raised, there was always ample organic manure.

Until I was grown I never heard of anyone having cancer.

In fact, at first cancer seemed to be coming from far off. For a long time if the subject of cancer came up, you could be sure cancer itself wasn't coming any nearer than to some congested place in the North, then to Atlanta, seventy-odd miles away, then to Macon, forty miles away, then to Monticello, twenty miles away.... The first inhabitants of our community to die of acknowledged cancer were almost celebrities, because of this 'foreign' disease. But now, twenty-odd years later, cancer has ceased to be viewed as a visitor and is feared instead as a resident. Even the children die of cancer now, which, at least in the beginning, seemed a disease of the old.

Most of the people I knew as farmers left the farms (they did not own the land and were unable to make a living working for the white people who did) to rent small apartments in the towns and cities. They ceased to have gardens, and when they did manage to grow a few things they used fertilizer from boxes and bottles, sometimes in improbable colors and consistencies, which they rightly suspected, but had no choice but to use. Gone were their chickens, cows, and pigs. Gone their organic manure.

To their credit, they questioned all that happened to them. Why must we leave the land? Why must we live in boxes with hardly enough space to breathe? (Of course, indoor plumbing seduced many a one.) Why must we buy all our food from the store? Why is the price of food so high – and it so tasteless? The collard greens bought in the supermarket, they said, 'tasted like water'.

The United States should have closed down and examined its every intention, institution, and law on the very first day a black woman observed that the collard greens tasted like water. Or when the first person of any color observed that store-bought tomatoes tasted more like unripened avocados than tomatoes.

The flavor of food is one of the clearest messages the Universe ever sends to human beings; and we have by now eaten poisoned warnings by the ton.

When I was a child growing up in middle Georgia in the forties and fifties, people still died of old age. Old age was actually a common cause of death. My parents inevitably visited dying persons over the long or short period of their decline; sometimes I went with them. Some years ago, as an adult, I accompanied my mother to visit a very old neighbor who was dying a few doors down the street, and though she was no longer living in the country, the country style lingered. People like my mother were visiting her constantly, bringing food, picking up and returning laundry, or simply stopping by to inquire how she was feeling and to chat. Her house, her linen, her skin all glowed with cleanliness. She lay propped against pillows so that by merely turning her head she could watch the postman approaching, friends and relatives arriving, and, most of all, the small children playing beside the street, often in her yard, the sound of their play a lively music.

Sitting in the dimly lit, spotless room, listening to the lengthy but warm-with-shared memories silences between my mother and Mrs. Davis was extraordinarily pleasant. Her white hair gleamed against her kissable black skin, and her bed was covered with one of the most intricately patterned quilts I'd ever seen – a companion to the dozen or more she'd stored in a closet, which, when I expressed an interest, she invited me to see.

I thought her dying one of the most reassuring events I'd ever witnessed. She was calm, she seemed ready, her affairs were in order. She was respected and loved. In short, Mrs. Davis was having an excellent death. A week later, when she had actually died, I felt this all the more because she had left, in me, the indelible knowledge that such a death is possible. And that cancer and nuclear annihilation are truly obscene alternatives. And surely, teaching this very vividly is one of the things an excellent death is supposed to do.

To die miserably of self-induced sickness is an aberration we take as normal; but it is crucial that we remember and teach our children that there are other ways.

For myself, for all of us, I want a death like Mrs. Davis's. One in which we will ripen and ripen further, as richly as fruit, and then fall slowly into the caring arms of our friends and other people we know. People who will remember the good days and the bad, the names of lovers and grandchildren, the time sorrow almost broke, the time loving friend-ship healed.

It must become a right of every person to die of old age. And if we secure this right for ourselves, we can, coincidentally, assure it for the planet. And that, as they say, will be excellence, which is, perhaps, only another name for health.

'Longing to die of old age' (1985) from Alice Walker, *Living by the Word*, London, Women's Press (1988), 33–6.

CHAPTER 14

Eco-feminism

L. Goodison, *Were the Greeks Green?*
A. Conway, *A mere dead mass*
E. Goldman, *Mother Earth*
D. Russell, *The soul of Russia and the body of America*
S. Griffin, *Matter*

Introduction

In local communities, unnumbered anonymous women took up specific causes. A housewife named Claire Dedrick, trying to stop a road in San Mateo country, California, in 1966, was told by an exasperated engineer: 'Get back in your kitchen, lady, and let me build my road!' The remark launched a career in conservation. Joining the local Sierra Club chapter, Dedrick went on to the club's board of directors and, in 1975, became California's Secretary for Resources.

(N. Cohen 1984: 344)

Eco-feminism argues that the struggle for ecological survival is intrinsically linked with the project of women's liberation. Eco-feminism in its various forms, looking back to the Suffragettes, instrumental in the campaign against cruise missiles and nuclear power, is an activist movement committed to social justice. Many eco-feminists would see themselves as would-be revolutionaries in an anarchist or socialist tradition. Capitalism and hierarchy stem from a patriarchal system of male rule. In its more mystical form, eco-feminism celebrates Gaia as a living force and looks back to a period variously situated in the early Bronze Age, the Neolithic or the Palaeolithic, a time when a supposedly matriarchal order existed in harmony with nature. Such an ancient society functioned without war or inequality.

Most controversially it has been argued that women are intrinsically more ecological than men. Such a supposition is often attacked by other feminists as biological, essentialist and hostile to true liberation. According to critics like Biehl (1988), a school of thought that labels women as Greener and nurturing may risk placing them back in the kitchen and nursery. Even identifying a distinct eco-feminist history can be seen as a sexist project, separating the intellectual and practical fruit of women's labour and consigning their efforts to a distinct pigeonhole. Every area of Green ideology and practice has received significant contributions from women. The Quakers, instrumental as peace and ecology activists in the twentieth century, founded at the time of the English civil war of the seventeenth century, were noted for the involvement of women in their activities. Romanticism at the turn of the nineteenth century, with its clear environmental sympathy, had important female exponents, including Mary Shelley, author of *Frankenstein*. Despite the misanthropy of Ruskin and the dubious outlook of Morris (in his utopian *News from Nowhere* women remain consigned to serving the communal meals), many of the 1880s generation of eco-socialists combined feminist radicalism with their ideas. Octavia Hill, founder of the National Trust, was perhaps the leading conservationist of her day. Women were active in the progressive conservation movement of Roosevelt's America at the start of the twentieth century. More recently ecological feminists have been involved in the peace movement and radical environmental groups such as Earth First!

Our first extract sheds tentative light on the concept of prehistoric matriarchy, with Lucy Goodison suggesting that archaeological evidence indicates 'a female oriented society in Crete'. Early Bronze Age Crete, far from worshipping a single Earth Mother deity, seems to have rejected all gods and goddesses, instead practising 'a religion centred on animals, plants and elements of the natural world'. Goodison questions 'the traditional identification of nature with the female principle' but does feel that this ancient society was female in orientation, non-hierarchical and peaceful.

Anne Conway (1631–79), a Quaker, opposed the reductionist philosophy of Descartes. For Merchant she represents one of many women active in the fields of philosophy and science who have been hidden from the male-inscribed historical record. As a vitalist for whom the entire universe was somehow alive, she seems to have supplied Liebniz with the holistic concept of the monad (Merchant 1980: 254).

Emma Goldman (1869–1940), the Russian-born anarchist, was inspired by a ride through the countryside, where she saw the fresh shoots of new plants breaking through the soil. She published the journal *Mother Earth*. An anti-state communist and feminist, she reminds us that whoever severs themself 'from Mother Earth goes into exile'.

Dora Russell, whose lifelong political activism overlapped with both Goldman and the Greenham women of the 1980s, can be seen as something of a bridge for eco-feminist thought. In this extract written around 1919 but unpublished until the 1980s, Russell, a passionate socialist, wonders whether the 1917 revolutions, given Russia's agrarian past, might provide a Green future for the Soviet Union, with industrialization carefully controlled by humanity. Although the USSR disappointed her, she continued a lifelong campaign for a socialism that served both women and nature.

Finally, Susan Griffin, a 1980s eco-feminist, in her long essay *Women and Nature*, writing in a style which is neither poetry nor prose but something of both, tracks ideas that have led to the exploitation of women and the Earth.

Lucy Goodison

Were the Greeks Green?

They say the Greeks had a word for it. From Oedipus to Electra, from Hercules to Pandora, ancient Greek myths, ideas and practices have been used to provide a pedigree for Western patriarchal values and preoccupations. In reaction, some feminists have claimed earlier traditions of matriarchy and a 'Goddess'. But long-neglected evidence from prehistoric Crete suggests something rather more surprising: a female-oriented society whose attitudes to the natural world had much in common with modern 'green' consciousness.

Traditionally archaeologists have looked to the Aegean Bronze Age (roughly 3000–1100 B.C.) for the first signs of Zeus and the Olympian gods. In his excavation of the Cretan 'palace' of Knossos, Sir Arthur Evans was interested in finding the first signs of the 'Cretan Zeus', but most of the pictures he found featured not men but women, in prominent positions. After 2000 years of the male Christian God, perhaps it is understandable that some women have found consolation in the idea that the earliest religions began with a 'Great Goddess', 'Mother Goddess', or 'Earth Goddess'. But after over a decade of research I didn't find enough evidence for any personified 'Goddess' in the Greek Early Bronze Age. The female figurines from this period are mixed with many male, unsexed and animal figurines which could equally be deities. Moreover, the female figurines are mostly small, have not been treated with particular respect, are sometimes paired (a monotheistic deity does not come in twos) and hardly ever hold a child – a serious problem for a 'Mother Goddess'. Instead we find numerous representations reflecting an interest in, and reverence for, the natural environment.

I believe that unquestioned assumptions have blinded people to evidence of a qualitatively different kind of religious attitude. One that originally did not focus on a male sky god, nor on his counterpart, a female earth goddess – nor indeed on any centralised or humanised image of divinity, nor on the patterns of male/female dualistic thought which we take for granted. Rather the scenes engraved on the tiny seals which these early Cretans used suggest a religion centred on animals, plants and elements of the natural world. In several cases humans are dressed in animal or bird costume and appear to take part in dancing or rituals with animals. The pictures suggest respect for the environment and the physical world, as figures reverentially touch a plant on an altar, or a goat, or celebrate the body through dance, or salute the sun.

Scholars accept that a sun deity figured large in the ancient religion of the Near East, and that the sun was important in the religious ideas of the Egyptians at this time, but they have perhaps regarded sun worship as too barbarian or primitive for the Greeks. Certainly the evidence for sun worship has been largely ignored. The pictures on the seals are backed up by architectural evidence: tholos tombs in Southern Crete face the rising sun, and most of the Cretan palace courtyards run North–South so that important rooms face east. What is particularly interesting is that it is usually women who are shown dancing to the sun, while from the nearby Cycladic islands some ritual vessels (nicknamed 'frying-pans') have a sun decorated on the belly of the vessel just above a female pubic triangle, suggesting that the sun was imagined to be female and was symbolically linked with female sexuality and the female belly.

Cave and mountain peak sanctuary cults (where offerings were sometimes placed in a chasm) suggest that earth was not an inferior partner to this cult of the heavenly bodies, but was sacred too. In fact there is nothing to suggest that these early people experienced our great divide between sky and earth, light and dark, nor the divide that we experience between spirit and matter, between the divine and the physical. 'Fragments of evidence (boats in graves, corpses buried in foetal position, circular East-facing tombs, rituals at cemeteries) cumulatively build up to suggest a circular scheme of belief which saw the sun as a creative and regenerative force. Its disappearance and reappearance were apparently linked with the rebirth of plant, animal and human life. Instead of dualism we see an integrated world view which located the divine in the physical world.

In the Early and Middle Bronze Age in Crete, such attitudes can be linked to a particular type of society. There isn't enough evidence to suggest a matriarchy, but tenuous and fragmentary evidence for social organisation points to a strong agricultural base; a way of life not strongly oriented towards warlike activity; some collectivity in the organisation both of the living community and of the burial of the dead. There is no clear evidence for hierarchies of wealth or power before the full development of the 'palaces'.

What happened over the following centuries, and how attitudes to women and nature changed as social structures changed and patriarchy became established, is another story. In the Late Bronze Age the development of hierarchies of land ownership and religious control has been linked to the influence of the militaristic Mycenaeans from the Greek mainland. In this period there are plenty of goddesses – and gods – and reverence for the natural world seems to wane. By the time of the 'Geometric' period (circa 900–700 B.C.), women were second-class citizens, the body was considered inferior to the divine, and 'Earth' had for the first time become exclusively seen as female. This is also when those patriarchal Greek myths appeared which, with their bright manly heroes and dark sinister females, serve as a pedigree for many misogynist psychological theories today. Attitudes towards the earth as an inferior element are reflected in the creation myth about the breach between Heaven and Earth as estranged husband and wife. The idea of a conflict between 'man' and 'nature' is beginning to crystallise.

Really changing the relationship between humans and the natural world means questioning the traditional identification of nature with a female principle alien to the male principle of intelligence and progress. The Cretan material reminds us that such symbolic identifications are arbitrary, and historically specific. While some strands of the Green movement celebrate nature and rehabilitate traditional 'female' qualities, this can be seen as another way of maintaining the same divisions. Perhaps it is more fruitful to acknowledge the strong (and often denied) connections between 'man' himself and the earth, between nature and culture, as reflected in the daily transactions of industrial production and the social infrastructure. To bridge the traditional intellectual separation between economics and ecology we need to drop some restricting stereotypes, including the tenacious notion that the earth is to society what woman is to man.

Reference

Moving Heaven to Earth; Sexuality, Spirituality and Social Change, by Lucy Goodison, is published by the Women's Press in London.

'Were the Greeks Green?', *Greening the Planet* 3 (1992), 24–5.

Anne Conway

A mere dead mass

1. The Philosophers (so called) of all Sects, have generally laid an ill Foundation to their Philosophy; and therefore the whole Structure must needs fall. 2. The Philosophy here treated on is not Cartesian. 3. Nor the Philosophy of *Hobbs* and *Spinosa*, (falsely so feigned,) but diametrically opposite to them. 4. That they who have attempted to refute *Hobbs* and *Spinosa*, have given them too much advantage. 5. This Philosophy is the strongest to refute *Hobbs* and *Spinosa*, but after another method. 6. We understand here quite another thing by Body and Matter, than *Hobbs* understood; and which *Hobbs*, and *Spinosa*, never saw, otherwise than in a Dream. 7. Life is as really and properly an Attribute of Body, as Figure. 8. Figure and Life are distinct, but not contrary Attributes of one and the same thing. 9. Mechanical Motion and Action or Perfection of Life, distinguishes Things.

1. From what hath been lately said, and from divers Reasons alledged, That Spirit and Body are originally in their first Substance but one and the same thing, it evidently appears that the Philosophers (so called) which have taught otherwise, whether Ancient or Modern, have generally erred and laid an ill Foundation in the very beginning, whence the whole House and superstructure is so feeble, and indeed so unprofitable, that the whole Edifice and Building must in time decay, from which absurd Foundation have arose very many gross and dangerous Errours, not only in Philosophy, but also in Divinity (so called) to the great damage of Mankind, hindrance of true Piety, and contempt of God's most Glorious Name, as will easily appear, as well from what hath been already said, as from what shall be said in this Chapter.

2. And none can Object, That all this Philosophy has no other than that of *des Cartes*, or *Hobbs* under a new Mask. For, First, as touching the *Cartesian* Philosophy, this saith that every Body is a mere dead Mass, not only void of all kind of Life and Sense, but utterly uncapable thereof to all Eternity; this grand Errour also is to be imputed to all those who affirm Body and Spirit to be contrary Things, and inconvertible one into another, so as to deny a Body all Life and Sense; which is quite contrary to the grounds of this our Philosophy. Wherefore it is so far from being a *Cartesian* Principle, under a new Mask, that it may be truly said it is *Anti-Cartesian*, in regard of their Fundamental Principles; although it cannot be denied that *Cartes* taught many excellent and ingenious Things concerning the Mechanical part of Natural Operations, and how all Natural Motions proceed according to Rules and Laws Mechanical, even as indeed Nature her self, *i.e.* the Creature, hath an excellent Mechanical Skill and Wisdom in it self, (given it from God, who is the Fountain of all Wisdom,) by which it operates: But yet in Nature, and her Operations, they are far more than merely Mechanical; and the same is not a mere Organical Body, like a Clock, wherein there is not a vital Principle of Motion; but a living Body, having Life and Sense, which Body is far more sublime than a mere Mechanism, or Mechanical Motion.

3. But, Secondly, as to what pertains to *Hobbs's* Opinion, this is yet more contrary to

this our Philosophy, than that of *Cartes*; for *Cartes* acknowledged God to be plainly Immaterial, and an Incorporeal Spirit. *Hobbs* affirms God himself to be Material and Corporeal; yea, nothing else but Matter and Body, and so confounds God and the Creatures in their Essences, and denies that there is any Essential Distinction between them. These and many more the worst of Consequences are the Dictates of *Hobbs's* Philosophy; to which may be added that of *Spinosa*; for this *Spinosa* also confounds God and the Creatures together, and makes but one Being of both; all which are diametrically opposite to the Philosophy here delivered by us.

The Principles of the most Ancient and Modern Philosophy (1690), The Hague: Nijhoff (1982), 221–2.

Emma Goldman

Mother Earth

There was a time when men imagined the Earth as the centre of the universe. The stars, large and small, they believed were created merely for their delectation. It was their vain conception that a supreme being, weary of solitude, had manufactured a giant toy and put them into possession of it.

When, however, the human mind was illumined by the torch-light of science, it came to understand that the Earth was but one of a myriad of stars floating in infinite space, a mere speck of dust.

Man issued from the womb of Mother Earth, but he knew it not, nor recognised her, to whom he owed his life. In his egotism he sought an explanation of himself in the infinite, and out of his efforts there arose the dreary doctrine that he was not related to the Earth, that she was but a temporary resting place for his scornful feet and that she held nothing for him but temptation to degrade himself. Interpreters and prophets of the infinite sprang into being, creating the 'Great Beyond' and proclaiming Heaven and Hell.... He was taught that Heaven, the refuge, was the very antithesis of Earth, which was the source of sin. To gain for himself a seat in Heaven, man devastated Earth. Yet she renewed herself, the good mother, and came again each spring, radiant with youthful beauty, beckoning her children to come to her bosom and partake of her bounty. But ever the air grew thick with mephitic darkness, ever a hollow voice was heard calling: 'Touch not the beautiful form of the Sorceress; she leads to Sin!'

But if the priests decried the Earth, there were others who found it a source of power and who took possession of it. Then it happened that the autocrats at the gates of heaven joined forces with the powers that had taken possession of the Earth; and humanity began its aimless, monotonous march. But the good mother sees the bleeding feet of her children, she hears their moans, and she is ever calling them that She is theirs....

Whoever severs himself from Mother Earth and her flowing sources of life goes into exile. A vast part of civilisation has ceased to feel the deep relation with our mother. How they hasten and fall over one another, the many thousands of the great cities; how they swallow their food, everlastingly counting the minutes with cold, hard faces; how they dwell packed together, close to one another, above and beneath, in dark gloomy stuffed holes, with dull hearts and insensitive heads, from lack of space and air. Economic necessity causes such hateful pressure. Why not economic stupidity? This seems a more appropriate name for it. Were it not for lack of understanding and knowledge, the necessity of escaping from the agony of an endless search for profit would make itself felt more keenly.

Must the Earth forever be arranged like an ocean steamer, with large luxurious rooms and luxurious food for a select few, and underneath the steerage, where the great mass can barely breathe from dirt and poisonous air? Neither unconquerable external nor internal necessity forces the human race to such life; that which keeps it in such a condition are ignorance and indifference.

Mother Earth Bulletin 1, 1 (1906–7), 1–2; 1, 2 (1906–7), 2.

Dora Russell

The soul of Russia and the body of America

It is not difficult to define what is meant by the body of America. As one writes the words, the imagination conjures up visions of sky-scrapers with swift elevators; vast factories where materials can be seen travelling fantastically on moving platforms to emerge at the exit as finished products; huge freight cars thundering their way from one busy town to another, immense liners ploughing the Atlantic; wide fields of cotton, great expanses of ripening corn. America stands, in fact, for the most complete example of the mechanism of industrial production on which the whole economic life of the West is based. It is an impressive mechanism, so impressive that quite three-quarters of those involved in, or in contact with, it forget that it is but a mechanism and nothing more. They come to imagine that this organisation of economic life, this speed, this comfort are in themselves civilisation and the goal of human endeavour, that all the best creative energy of man should be turned to developing resources producing goods, inventing processes to speed up production.

They endow this machine with a soul and a message which is to be carried to the uttermost parts of the earth, to be taught if need be, by bullying, or at the point of the sword. To those who, despite every effort to the contrary, cannot bring themselves to accept such a primitive notion of civilisation, this machine worship is as horrible and superstitious as the adoration of the savage for his painted block of wood or stone. There have been, in the past, many of these dissenters. And one sees them now, in America, enquiring distressfully what is the matter with their country, feeling dimly that the trouble lies in her barrenness of ideals and emptiness of soul, and, looking round from one party to another, and one class to another, seeking a possible source of regeneration.

In Europe, too, idealists are trying to find some motive for building prosperity anew, and the disgust and despair in which the war has left them are but heightened when they look across and see in America the image of what they may become, of what America is capable of making of the whole world. They see this excellent body, this shell of a state, and the soul of man walking mournfully through it, as through a wilderness, seeking an oasis where it may perchance rest for a moment, not hoping to find a home.

It is not from America that regeneration can come. There is every sign that her people, like the industrial peoples of Europe, will first seek relief from the intolerable mechanical burden of their lives in the worn-out pastime of imperial conquest. Yet all that America could give to a subservient world would be her body, her industrial efficiency, a valuable gift in days gone by, and still needed in the present and the future, but not enough. America can give us no new ideals, and it is for new ideals that the whole world, from the east Atlantic to the west Pacific, is hungry. Thinking Europe has become conscious at last that it cannot live with the industrial machine unless new ideals can be found to control and govern it. In China, also, the question on the lips of all intelligent people is: 'Since it seems we must follow in the path of the industrial nations, how shall we do so without becoming as horrible and degraded as they?'

One nation in the world has set out to answer that question in practice, and that is Russia. For this reason the most cynical have turned to her in joyful surprise; even her bitterest and blindest opponents are conscious that she has found something new which she is trying to expound to the world and, while they do their utmost to destroy her in the act of realisation of her ideals, they yet have a sneaking hope that they may not succeed. So desperate has the need for hope become in our blackened and ruined world. . . .

Often while in Russia and since returning, I have wondered whether we are right or they. We, who have conceived of communism as budding and blossoming like a flower on the sturdy plant of competent and organised industry, or they, who see it as a whirling heart of fire that must consume ancient evils and then, cooling, transmute itself into the crust of material expression, creating industrialism anew, a thing, it may be, of undreamed-of power and beauty. To us, tutored in determinism, economic circumstances is the decisive factor in politics. We think of the industrial machine as having an irresistible momentum, we imagine Russia in its grip, changing ideals and character, assimilating rapidly to the industrial nations of the West. But when we do so, we forget how far the industrial system, as *we* conceive it, is the product of the thought of our past, how it perpetuates old prejudices, how it bears, like every thought or institution in the world, the unmistakable stamp of its origin and date.

Two visions came repeatedly before my eyes. In the one the machine in America grew increasingly rapacious and cruel, while in Russia it triumphed over human forces, and Europe and Asia were sucked into its maw. There were long hours of mechanical slavery, black and ugly factories, fatuous towns and futile luxuries. Thought and art were dead; the populations petulant and trivial.

In the other the spirit of communism in Russia had leapt like a great wave to meet the West, and Western science and skill – its twin brother – had reared its head and sprung to the meeting with an exultant roar. So they met at last, soul and body, and went springing skywards in a clear, green pyramid of joy. The filth of factories and the grime of poverty were washed away and everywhere there emerged a new and smiling world. Human life was restored to harmony; men were no longer cramped and twisted to serve as wheels and cogs; they found that leisure to savour the whole life of man is better than empty luxury that cogs cannot enjoy. The power of the machine was broken for ever; it served instead of commanding, and everywhere the bright roofs of lovely hamlets, the spacious factories, the grassy tree-girt spaces where children and students met to chatter and play, and workers to dance and sing after their easy labours; the quiet arbours where the artist would seek loneliness to brood, or the men of science peace for arduous discussion or complicated thought – all these testified to what life might be, not for the few, but for all, if the spirit of man in justice and humanity would but conquer and yoke the mechanical monster to his will.

Our Western industrial body can give birth to this vision, but can it unfold the spirit that could achieve its realisation? But I am confident that communism, cutting out from the industrial system the motives of profit and exploitation, and administering it in terms of humanity and justice, could so transform industrialism as to make of it a thing of beauty, not of terror.

Text exactly as written in 1920–1, when the author was twenty-six. From *The Religion of the Machine Age*, London: Routledge, ix–xv.

Susan Griffin

Matter

It is decided that matter is transitory and illusory like the shadows on a wall cast by firelight; that we dwell in a cave, in the cave of our flesh, which is also matter, also illusory; it is decided that what is real is outside the cave, in a light brighter than we can imagine, that matter traps us in darkness. That the idea of matter existed before matter and is more perfect, ideal.

Sic transit, how quickly pass, *gloria mundi*, the glories of this world, it is said.

Matter is transitory and illusory, it is said. This world is an allegory for the next. The moon is an image of the Church, which reflects Divine Light. The wind is an image of the Spirit. The sapphire resembles the number eleven, which has transgressed ten, the number of the commandments. Therefore the number eleven stands for sin.

It is decided that matter is passive and inert, and that all motion originates from outside matter.

That the soul is the cause of all movement in matter and that the soul was created by God: that all other movement proceeds from violent contact with other moving matter, which was first moved by God. That the spheres in perpetual movement are moved by the winds of heaven, which are moved by God, that all movement proceeds from God.

That matter is only a potential for form or a potential for movement.

It is decided that the nature of woman is passive, that she is a vessel waiting to be filled.

It is decided that the existence of God can be proved by reason and that reason exists to apprehend God and Nature.

God is unchangeable, it is said. *Logos* is a quality of God created in many by God and it is eternal. The soul existed before the body and will live after it.

'And I do not know how long anything I touch by a bodily sense will exist,' the words of a saint read, 'as, for instance, this sky and this land, and whatever other bodies I perceive in them. But seven and three are ten and not only now but always ... therefore ... this incorruptible truth of numbers is common to me and anyone at all who reasons.'

And it is stated elsewhere that Genesis cannot be understood without a mastery of mathematics.

'He who does not know mathematics cannot know any of the other sciences,' it is said again, and it is decided that all truth can be found in mathematics, that the true explanation is mathematics and fact merely evidence.

That there are three degrees of abstraction, each leading to higher truths. The scientist peels away uniqueness, revealing category; the mathematician peels away sensual fact, revealing number; the metaphysician peels away even number and reveals the fruit of pure being.

It is put forward that science might be able to prolong life for longer periods than might be accomplished by nature. And it is predicted:

> that machines for navigation can
> be made without rowers so that the
> largest ships on rivers or seas will
> by a single man be propelled with
> greater velocity than if they were
> full of men
> that cars can be made to move with
> out the aid of animals at an un
> believable rapidity
> that flying machines can be con
> structed
> that such things can be
> made without limit.

It is decided that vision takes place because of a ray of light emanating from the eye to the thing perceived.

It is decided that God is primordial light, shining in the darkness of first matter, giving it substantial being. It is decided that geometrical optics holds the key to all understanding.

It is said that the waters of the firmament separate the corporeal from the spiritual creation.

That the space above is infinite, indivisible, immutable, and is the immensity of God.

That the earth is a central sphere surrounded by concentric zones, perfect circles of air, ether and fire, containing the stars, the sun and the planets, all kept in motion by the winds of heaven. That heaven is beyond the zone of fire and that Hell is within the sphere of the earth. That Hell is beneath our feet.

It is stated that all bodies have a natural place, the heavy bodies tending toward the earth, the lighter toward the heavens.

And what is sublunary is decaying and corruptible. The earth 'is so depraved and broken in all kinds of vice and abominations that it seemeth to be a place that hath received all the filthiness and purgings of all other worlds and ages,' it is said.

And the air below the moon is thick and dirty, while the air above 'shineth night and day of resplendour perpetual,' it is said.

And it is decided that the angels live above the moon and aid God in the movement of celestial spheres. 'The good angels,' it is said, 'hold cheap all the knowledge of material and temporal matters which inflates the demon with pride.'

And the demon resides in the earth, it is decided, in Hell, under our feet.

It is observed that women are closer to the earth.

That women lead to man's corruption. Women are 'the Devil's Gateway,' it is said.

That regarding the understanding of spiritual things, women have a different nature than men, it is observed, and it is stated that women are 'intellectually like children'. That women are feebler of body and mind than men, it is said: 'Frailty, thy name is woman.'

And it is stated that 'the word woman is used to mean the lust of the flesh'.

That men are moved to carnal lust when they hear or see woman, whose face is a burning wind, whose voice is a hissing serpent.

It is decided that in birth the female provides the matter (the menstruum, the yolk) and that the male provides the form which is immaterial, and that out of this union is born the embryo.

And it is written in the scripture that out of Adam who was the first man was taken Eve, and because she was born of man he also named her: 'She shall be called Woman.'

And it is written in the bestiary that the cubs of the Lioness are born dead but on the third day the Lion breathes between their eyes and they wake to life.

It is decided that Vital Heat is the source of all vital activity, that this heat emanates from God to the male of the species, and that this vital heat informs the form of the species with maleness, whereas the female is too cold to effect this change.

It is decided also that all monstrosities of birth come from a defect in the matter provided by the female, which resists the male effort to determine form.

It is decided that Vital Heat is included in semen, that it is the natural principle in the spirit and is analogous to that element in the stars.

It is decided that the Vital Heat of the sun causes spontaneous generation.

The discovery is made that the sun and not the earth is the center of the universe. And the one who discovers this writes:

'In the middle of all sits Sun enthroned. In this most beautiful temple could we place this luminary in any better position from which he can illuminate the whole at once? He is rightly called the Lamp, the Mind, the ruler of the Universe; Hermes Trismegistus names him the visible God, Sophocles' Electra calls him the All-Seeing. So the Sun sits as upon a royal throne ruling his children the planets which circle round him.... Meanwhile the earth conceives by the Sun, and becomes pregnant with an annual rebirth.'

And it is decided that the Sun is God the Father, the stars God the Son, and the ethereal medium the Holy Ghost.

Mutability on the earth, it is said, came to the Garden of Eden after the Fall. That before the Fall there was immortal bliss on earth, but that after the Fall 'all things decay in time and to their end do draw'.

That the face of the earth is a record of man's sin. That the height of mountains, the depth of valleys, the sites of great boulders, craters, seas, bodies of land, lakes and rivers, the shapes of rocks, cliffs, all were formed by the deluge, which was God's punishment for sin.

'The world is the Devil and the Devil is the world,' it is said.

And of the fact that women are the Devil's Gateway it is observed that sin and afterwards death came into the world because Eve consorted with the devil in the body of a serpent.

That the power of the devil lies in the privy parts of men.

That women act as the devil's agent and use flesh as bait.

That women under the power of the devil meet with him secretly, in the woods (in the wilderness), at night. That they kiss him on the anus. That they offer him pitch-black candles, which he lights with a fart. That they anoint themselves with his urine. That they dance back to back together and feast on food that would nauseate 'the most ravenously hungry stomach'. That a mass is held, with a naked woman's body as an altar, feces, urine and menstrual blood upon her ass. That the devil copulates with all the women in this orgy, in this ritual.

That these women are witches.

That 'Lucifer before his Fall, as an archangel, was a clear body, composed of the purest and brightest air, but that after his Fall he was veiled with a grosser substance and took a new form of dark and thick air.'

Women and Nature, New York: Harper & Row (1978), 5–10.

CHAPTER 15

Spiritual awakenings

E. F. Schumacher, Buddhist economics
Hildegard of Bingen, The Book of Divine Works
J. Bauthumley, The light and dark sides of God
Starhawk, Witchcraft as Goddess religion
S. H. Nasr, Islam and the Earth

Introduction

> The vulgar boast of the modern technologist that man has conquered nature
> has roots in the Western religious tradition, which affirms that God installed
> man as the boss, to whom Nature was to bring tribute. The Greeks knew better
> than the Jews or Christians. They knew that hubris towards nature was as
> much of a sin as hubris towards fellow men. Xerxes is punished not only for
> having attacked the Greeks, but also for having outraged nature in the affair of
> bridging the Hellespont. But for an ethical system that includes animate and
> inanimate Nature as well as man, one must go to Chinese Taoism. . . .
>
> (Huxley 1969: 578–9)

A recurring theme in virtually all religions has been the relationship between humanity, nature and the sacred. Shamanism and other forms of paganism, which hold that the Earth is sacred, literally alive and inhabited by hosts of nature spirits or Devas, have been seen as the original Earth religion. Conversely, monotheistic Judaeo-Christianity, with its rejection of the Earth mother in favour of a male God residing above us, has been criticized as anti-ecological. Christianity, and to a lesser extent Judaism and Islam, have been blamed for separating humanity and nature as well as promoting the idea that nature was created for humanity to exploit. The Judaeo-Christian concept of original sin may have given rise to the doctrine that, far from being sacred and worthy of reverence, the Earth is a fallen realm that deserves little care. Eastern religions, including Buddhism, Taoism and Hinduism, with their opposition to materialist greed, their love of nature, contemplation and meditation, along with ethical vegetarianism, have been promoted as alternatives. Christians and Moslems have argued in response that humanity has been given the role of a steward or shepherd by God. More recently the Catholic theologian Matthew Fox (1987) has countered the concept of original sin with that of original blessing, stressing that the Earth is sacred. Many Eastern religions, including Hinduism, unlike Christianity, hold that the Earth and indeed all material 'reality' is illusory. This attitude, Christians have claimed, is hardly reverential towards nature.

Many Greens have argued that political and social problems can be solved only by spiritual change. Economic transformation alone is not enough. A new ethic, which rejects consumerism and links us to nature is necessary. Concepts of philosophical holism, non-violence, respect for other species and eco-feminism, etc., are embraced by Green spirituality. Such a view linking religious ethics to political economy is captured by the author of *Small is Beautiful*, F. E. Schumacher (1911–73). His outstanding contribution to the creation of a modern Green movement was inspired directly by his experience in Burma, where as a development economist he came to believe that Buddhist concepts of economics were more relevant and sophisticated than those of the West (Wood 1984).

Hildegard of Bingen (1098–1179), the twelfth-century German abbess, is often classed with St Francis as an environmentally aware Christian mystic. In her view of nature, by contrast with that of St Augustine, God created sacred wilderness. She not only celebrates the spirit but also argues that the flesh is holy. For her, God is literally Green. Her almost pagan Christianity comes close to the pantheism of our next extract, taken from the seventeenth-century essay *The Light and Dark Side of God*. In this tract a heretical Protestant Ranter proclaims that all of nature is God, in a rare example of what may be a hidden Western pagan tradition. Jacob Bauthumley, who like Hildegard saw God in 'every green thing', was made to suffer for his belief. Court-martialled for blasphemy, he was expelled from his regiment on 14 March 1650; his sword was broken over his head, his pamphlet burned and a red hot iron pushed through the flesh of his tongue (Davis 1986: 44).

According to Starhawk, an American anti-nuclear campaigner and feminist, witchcraft is the religion of Gaia. She defends witchcraft as a legitimate religion, outlines its main themes and shows it to be ecologically orientated. Both Starhawk and the Ranters combine paganism with political sensibilities biased towards equality and social justice.

Finally the Islamic scholar Syed Hossein Nasr, in *Man and Nature*, examines the ecological credentials of Islam. For Nasr, this religion, whose colour is green, provides a bridge between the nature-orientated outlook of the Eastern religions and the more human-centred Judaism and Christianity. He argues that spiritual values have regulated Islamic economics and technological development, preventing the destruction inherent in Western development. He also perceives nature as a central metaphor in Islam of all that is sacred. Yet Allah still places humanity as Caliph over nature, not as a humble equal with other species. Ultimately, for Nasr, while nature may be used as a means of reaching the sacred, as in Mohammed's meditations in the desert, the sacred remains above and beyond Earth-bound nature.

Green spirituality extends, of course, beyond any concept of religion. Green authors, including Bahro, Porritt and Leiss, emphasize the creative nature of humanity, its need to link with nature and oppose the commoditization of human needs in a capitalist society. Such sentiments can be found in the work of Tawney (who influenced Porritt 1984), within the Frankfurt school (see Chapter Eighteen) and also, it may be argued, in the 1844 manuscripts of Marx (Fromm 1961). In *To Have or to Be?* (Erich Fromm) is perhaps the most obvious exponent of such a Green spirituality, which draws upon earlier spiritual sources without worshipping any God or Goddess.

Fritz Schumacher

Buddhist economics

While the materialist is mainly interested in goods, the Buddhist is mainly interested in liberation. But Buddhism is 'The Middle Way' and therefore in no way antagonistic to physical well-being. It is not wealth that stands in the way of liberation but the attachment to wealth; not the enjoyment of pleasurable things but the craving for them. The keynote of Buddhist economics, therefore, is simplicity and non-violence. From an economist's point of view, the marvel of the Buddhist way of life is the utter rationality of its pattern – amazingly small means leading to extraordinarily satisfactory results.

For the modern economist this is very difficult to understand. He is used to measuring the 'standard of living' by the amount of annual consumption, assuming all the time that a man who consumes more is 'better off' than a man who consumes less. A Buddhist economist would consider this approach excessively irrational; since consumption is merely a means to human well-being, the aim should be to obtain the maximum of well-being with the minimum of consumption. Thus, if the purpose of clothing is a certain amount of temperature comfort and an attractive appearance, the task is to attain this purpose with the smallest possible effort, that is, with the smallest annual destruction of cloth and with the help of designs that involve the smallest possible input of toil. The less toil there is, the more time and strength is left for artistic creativity. It would be highly uneconomic, for instance, to go in for complicated tailoring, like the modern West, when a much more beautiful effect can be achieved by the skilful draping of uncut material. It would be the height of folly to make material so that it should wear out quickly and the height of barbarity to make anything ugly, shabby or mean. What has just been said about clothing applies equally to all other human requirements. The ownership and the consumption of goods is a means to an end, and Buddhist economics is the systematic study of how to attain given ends with the minimum means.

Modern economics, on the other hand, considers consumption to be the sole end and purpose of all economic activity, taking factors of production – land, labour, and capital – as the means. The former, in short, tries to maximise human satisfactions by the optimal pattern of consumption, while the latter tries to maximise consumption by the optimal pattern of productive effort. It is easy to see that the effort needed to sustain a way of life which seeks to attain the optimal pattern of consumption is likely to be much smaller than the effort needed to sustain a drive for maximum consumption. We need not be surprised, therefore, that the pressure and strain of living is very much less in, say, Burma than it is in the United States, in spite of the fact that the amount of labour-saving machinery used in the former country is only a minute fraction of the amount used in the latter.

Simplicity and non-violence are obviously closely related. The optimal pattern of consumption, producing a high degree of human satisfaction by means of a relatively low rate of consumption, allows people to live without great pressure and strain and to fulfil the primary injunction of Buddhist teaching: 'Cease to do evil; try to do good.' As physical resources are everywhere limited, people satisfying their needs by means of a modest use of resources are obviously less likely to be at each other's throats than people depending upon a high rate of use. Equally, people who live in highly self-sufficient local communities are less likely to get involved in large-scale violence than people whose existence depends on worldwide systems of trade.

From the point of view of Buddhist economics, therefore, production from local resources for local needs is the most rational way of economic life, while dependence on imports from afar and the consequent need to produce for export to unknown and distant people is highly uneconomic and justifiable only in exceptional cases and on a small scale. Just as the modern economist would admit that a high rate of consumption of transport services between a man's home and his place of work signifies a misfortune and not a high standard of life, so the Buddhist economist would hold that to satisfy human wants from faraway sources rather than from sources nearby signifies failure rather than success.... Another striking difference between modern economics and Buddhist economics arises over the use of natural resources.... The teaching of the Buddha ... enjoins a reverent and non-violent attitude not only to all sentient beings but also, with great emphasis, to trees. Every follower of the Buddha ought to plant a tree every few years and look after it until it is safely established, and the Buddhist economist can demonstrate without great difficulty that the universal observation of this rule would result in a high rate of genuine economic development independent of any foreign aid. Much of the economic decay of south-east Asia (as many other parts of the world) is undoubtedly due to a heedless and shameful neglect of trees.

Modern economics does not distinguish between renewable and non-renewable materials, as its very method is to equalise and quantify everything by means of a money price.... From a Buddhist point of view, of course, this will not do; the essential difference between non-renewable fuels like coal and oil on the one hand and renewable fuels like wood and water-power on the other cannot simply be overlooked. Non-renewable goods must be used only if they are indispensable, and then only with the greatest care and the most meticulous concern for conservation. To use them heedlessly or extravagantly is an act of violence, and while complete non-violence may not be attainable on this earth, there is nonetheless an ineluctable duty on man to aim at the ideal of non-violence in all he does.

Just as a modern European economist would not consider it a great economic achievement if all European art treasures were sold to America at attractive prices, so the Buddhist economist would insist that a population basing its economic life on non-renewable fuels is living parasitically, on capital instead of income. Such a way of life could have no permanence and could therefore be justified only as a purely temporary expedient. As the world's resources of non-renewable fuels – coal, oil and natural gas – are exceedingly unevenly distributed over the globe and undoubtedly limited in quantity, it is clear that their exploitation at an ever-increasing rate is an act of violence between men.

Small is Beautiful, London: Abacus (1974), 47–50.

195

Hildegard of Bingen

The Book of Divine Works

Hildegard's commission

For five years I had been troubled by true and wonderful visions. For a true vision of the unfailing light had shown me (in my great ignorance) the diversity of various ways of life. In the sixth year (which marked the beginning of the present visions), when I was sixty-five years of age, I saw a vision of such mystery and power that I trembled all over and – because of the frailty of my body – began to sicken. It was only after seven years that I finally finished writing down this vision. And so, in the year of our Lord's incarnation, 1163, when the apostolic throne was still being oppressed by the Roman Emperor, Frederick, a voice came to me from heaven, saying:

> O poor little figure of a woman; you, who are the daughter of many troubles, plagued by a grave multitude of bodily infirmities, yet steeped, nonetheless, in the vastness of God's mysteries – commit to permanent record for the benefit of humankind, what you see with your inner eyes and perceive with the inner ears of your soul so that, through these things, people may come to know their Creator and not recoil from worshipping him with the reverence due to him. And so, write these things, not according to your heart but according to my witness – for I am Life without beginning or end. These things were not devised by you, nor were they previously considered by anyone else; but they were pre-ordained by me before the beginning of the world. For just as I had foreknowledge of man before he was made, so too I foresaw all that he would need.

And so I, a poor and feeble little figure of a woman, set my hand to the task of writing – though I was worn down by so many illnesses, and trembling. All this was witnessed by that man [Volmar] whom (as I explained in my earlier visions) I had sought and found in secret, as well as by that girl [Richardis] whom I mentioned in the same context.

While I was doing this, I looked up at the true and living light to see what I ought to write. For everything which I had written since the beginning of my visions (or which I came to understand afterwards) I saw with the inner eyes of my spirit and heard with my inner ears, in heavenly mysteries, fully awake in body and mind – and not in dreams, nor in ecstasy, as I explained in my previous visions. Nor (as truth is my witness) did I produce anything from the faculty of the human sense, but only set down those things which I perceived in heavenly mysteries.

And again I heard a voice from heaven instructing me thus; and it said: 'Write in this way, just as I tell you.'

Spiritual awakenings

The source of all being

'I, the highest and fiery power, have kindled every living spark and I have breathed out nothing that can die. But I determine how things are – I have regulated the circuit of the heavens by flying around its revolving track with my upper wings – that is to say, with Wisdom. But I am also the fiery life of the divine essence – I flame above the beauty of the fields; I shine in the waters; in the sun, the moon and the stars, I burn. And by means of the airy wind, I stir everything into quickness with a certain invisible life which sustains all. For the air lives in its green power and its blossoming; the waters flow as if they were alive. . . .

And how could God be known to be life, except through the living things which glorify him, since the things that praise his glory have proceeded from him? For this reason, he placed the living and burning sparks to brighten his face. These sparks see that he has neither beginning nor end and (unable to have their fill of gazing at him), they look eagerly upon him without satiety, with a zeal that can never diminish. But how could he, who is alone immortal, be known if the angels did not gaze upon him in this way? If he did not have those sparks, how could his full glory be apparent? And how could he be known to be eternal, if no brightness proceeded from him? For there is nothing in creation that does not have some radiance – either greenness or seeds or flowers, or beauty – otherwise it would not be part of creation. For if God were not able to make all things, where would be his power?

The work of the soul in the body

The soul assists the flesh and the flesh, the soul. For every single work is perfected through soul as well as flesh, so that the soul is revived by doing good and holy works with the flesh. But the flesh is often irked when co-operating with the soul, and so the soul stoops to the level of the flesh and allows it to take delight in some deed, just as a mother causes her weeping child to laugh. And in this way, the flesh performs some good works with the soul, but mixed together with certain sins which the soul tolerates so that the flesh is not oppressed. For just as the flesh lives through the soul, so too, the soul is revived by doing good works with the flesh, because the soul has been stationed inside the work of the Lord's hands. In the same way that the sun, overcoming night, climbs until the middle of the day, so man, too, rises up, by avoiding corrupt deeds. And just as the sun declines in the afternoon, so too, the soul makes accord with the flesh. And as the moon is rekindled by the sun so that it does not disappear, so the flesh of man is sustained by the powers of the soul, so that it does not go to ruin.

The goodness of all created things

God's works are so secured by an all-encompassing plenitude, that no created thing is imperfect. It lacks nothing in its nature, possessing in itself the fullness of all perfection and utility.

And so all things which came forth through Wisdom, remain in her like a most pure and elegant adornment, and they shine with the most splendid radiance of their individual essence.

And when fulfilling the precepts of God's commandments, man, too, is the sweet and dazzling robe of Wisdom. He serves as her green garment through his good intentions and the green vigour of works adorned with virtues of many kinds. He is an ornament to her ears when he turns away from hearing evil whispers; a protection for her breast, when he rejects forbidden desires. His bravery gives glory to her arms, too, when he defends himself against sin. For all of these things arise from the purity of faith, adorned with the profound gifts of the Holy Spirit and the most just writings of the Doctors of the Church, when man has perfected them in faith through good works.

Hildegard of Bingen: an Anthology, London: SPCK (1990), 90–1, 96–7, 100.

Jacob Bauthumley

The light and dark sides of God

Nay, I see that God is in all Creatures, Man and Beast, Fish and Fowle, and every green thing, from the highest Cedar to the Ivey on the wall; and that God is the life and being of them all, and that God doth really dwell, and if you will personally; if he may admit so low an expression in them all, and hath his Being no where else out of the Creatures.

Further, I see that all the Beings in the World are but that our Being, and so he may well be said, to be every where as he is, and so I cannot exclude him from Man or Beast, or any other Creature: Every Creature and thing having that Being living in it, and there is no difference betwixt Man and Beast; but as Man carries a more lively Image of the divine Being then any other Creature: For I see the Power, Wisdom, and Glory of God in one, as well as another, onely in that Creature called Man, God appears more gloriously in then the rest.

And truly, I find by experience, the grand reason why I have, and many others do now use set times of prayer, and run to formall duties, and other outward and low services of God: the reason hath been, and is, because men look upon a God as being without them and remote from them at a great distance, as if he were locally in Heaven, and sitting there onely, and would not let down any blessing or good things, but by such and such a way and meanes.

But Lord, how carnall was I thus to fancie thee? Nay I am confident, that there is never a man under the Sun that lookes upon God in such a forme; but must be a grosse Idolator, and fancie some corporall shape of him, though they may call it spiritual.

Did men see that that God was in them, and framing all their thoughts, and working all their works, and that he was with them in conditions: what carnall spirit would reach out to that by an outward way, which spiritually is in him, and which he stands really possest of? and which divine wisdom sees the best, and that things can be no otherwise with him. I shall speak my own experience herein, that I have made God mutable as my self, and therefore as things and conditions have changed, I thought that God was angry or pleased, and to have faln a humbling my self; or otherwise in thankfulness, never looking or considering that God is one intire perfect and immutable Being, and that all things were according to the Councel of his own will, and did serve the designe of his own glory: but thought that my sins or holy walking did cause him to alter his purpose of good or evill to me.

But now I cannot looke upon any condition or action, but methinks there appears a sweet concurreance of the supreame will in it; nothing comes short of it, or goes beyond it, nor any man shall doe or be any thing, but what shall fall in a sweet compliance with it; It being the wombe wherein all things are conceived, and in which all creatures were formed and brought fourth.

Yea further, there is not the least Flower or Herbe in the Field but ther is the Divine being by which it is, that which it is; and as that departs out of it, so it comes to nothing, and so it is to day clothed by God, & to morrow cast into the Oven: when God ceases to live in it then it comes to nothing, and so all the visible Creatures are lively resemblances of the Divine being. But if this be so, some may say: Then look how many Creatures there are in the world, there is so many Gods, and then they dye and perrish, then must God also die with them, which can be no lesse then blasphemy to affirm.

To which I answer, and it is apparent to me, that all the Creatures in the world; they are not so many distinct Beings, but they are but one intire Being, though they be distinguished in respect of their formes; yet their Being is but one and the same Being, made out in so many formes of flesh, as Men and Beast, Fish and Fowle, Trees and Herbes: For though these two Trees and Herbes have not the life so sensibly or lively; yet it is certain there is a Life and Being in them, by which they grow to that maturity and perfection, that they become serviceable for the use of Man, as other Creatures are; and yet I must not exclude God from them; for as God is pleased to dwel in flesh, and to dwel with and in man, yet is he not flesh, nor doth the flesh partake of the divine Being. Onely this, God is pleased to live in flesh, and as the Scripture saith, he is made flesh; and he appeares in severall formes of flesh, in the forme of Man and Beast, and other Creatures . . .

The Light and Dark Sides of God (1658) in N. Smith, ed., *A Collection of Ranter Writings from the Seventeenth Century*, London: Junction Books (1983), 232–3.

Starhawk

Witchcraft as Goddess religion

The symbol of the Goddess conveys the spiritual power both to challenge systems of oppression and to create new, life-oriented cultures.

Modern Witchcraft is a rich kaleidoscope of traditions and orientations. Covens, the small, closely knit groups that form the congregations of Witchcraft, are autonomous; there is no central authority that determines liturgy or rites. Some covens follow practices that have been handed down in an unbroken line since before the Burning Times. Others derive their rituals from leaders of modern revivals of the Craft – the two whose followers are most widespread are Gerald Gardner and Alex Sanders, both British. Feminist covens are probably the fastest-growing arm of the Craft. Many are Dianic, a sect of Witchcraft that gives far more prominence to the female principle than the male. Other covens are openly eclectic, creating their own traditions from many sources. My own covens are based on the Faery Tradition, which goes back to the Little People of Stone Age Britain, but we believe in creating our own rituals, which reflect our needs and insights of today. In Witchcraft, a chant is not necessarily better because it is older. The Goddess is continually revealing Herself, and each of us is potentially capable of writing our own liturgy.

In spite of diversity, there are ethics and values that are common to all traditions of Witchcraft. They are based on the concept of the Goddess as immanent in the world and in all forms of life, including human beings.

Theologians familiar with Judeo-Christian concepts sometimes have trouble understanding how a religion such as Witchcraft can develop a system of ethics and a concept of justice. If there is no split between spirit and nature, no concept of sin, no covenant or commandments against which one can sin, how can people be ethical? By what standards can they judge their actions, when the external judge is removed from his place as ruler of the cosmos? And if the Goddess is immanent in the world, why work for change or strive toward an ideal? Why not bask in the perfection of divinity?

Love for life in all its forms is the basic ethic of Witchcraft. Witches are bound to honor and respect all living things, and to serve the life-force. While the Craft recognizes that life feeds on life and that we must kill in order to survive, life is never taken needlessly, never squandered or wasted. Serving the life-force means working to preserve the diversity of natural life, to prevent the poisoning of the environment and the destruction of species.

The world is the manifestation of the Goddess, but nothing in that concept need foster passivity. Many Eastern religions encourage quietism not because they believe the divine is truly immanent, but because they believe She/He is not. For them, the world is Maya, Illusion, masking the perfection of the Divine Reality. What happens in such a world is not really important; it is only a shadow play obscuring the Infinite Light. In Witchcraft, however, what happens in the world is vitally important. The Goddess is immanent, but She needs human help to realize Her fullest beauty. The harmonious balance of plant/animal/human/divine awareness is not automatic; it must constantly be renewed, and this is the true function of Craft rituals. Inner work, spiritual work, is most effective when it proceeds hand in hand with outer work. Meditation on the balance of nature might be considered a spiritual act in Witchcraft, but not as much as would cleaning up garbage left at a campsite or marching to protest an unsafe nuclear plant.

Witches do not see justice as administered by some external authority, based on a written code or set of rules imposed from without. Instead, justice is an inner sense that each act brings about consequences that must be faced responsibly. The Craft does not foster guilt, the stern, admonishing, self-hating inner voice that cripples action. Instead, it demands responsibility. 'What you send, returns three times over' is the saying – an amplified version of 'Do unto others as you would have them do unto you'. For example, a Witch does not steal, not because of an admonition in a sacred book, but because the threefold harm far outweighs any small material gain. Stealing diminishes the thief's self-respect and sense of honor; it is an admission that one is incapable of providing honestly for one's own needs and desires. Stealing creates a climate of suspicion and fear, in which even thieves have to live. And, because we are all linked in the same social fabric, those who steal also pay higher prices for groceries, insurance, taxes. Witchcraft strongly imparts the view that all things are interdependent and interrelated and, therefore, mutually responsible. An act that harms anyone harms us all.

Honor is a guiding principle in the Craft. This is not a 'macho' need to take offense at imagined slights against one's virility – it is an inner sense of pride and self-respect. The Goddess is honored in oneself, and in others. Women, who embody the Goddess, are respected, not placed on pedestals or etherealized, but valued for all their human qualities. The self, one's individuality and unique way of being in the world, is highly valued. The Goddess, like nature, loves diversity. Oneness is attained not through losing the self, but through realizing it fully. 'Honor the Goddess in yourself, celebrate your self, and you will see that Self is everywhere.'

In Witchcraft, 'All acts of love and pleasure are My rituals.' Sexuality, as a direct expression of the life-force, is seen as numinous and sacred. It can be expressed freely, so long as the guiding principle is love. Marriage is a deep commitment, a magical, spiritual, and psychic bond. But it is only one possibility out of many for loving, sexual expression.

Misuse of sexuality, however, is heinous. Rape, for example, is an intolerable crime because it dishonors the life-force by turning sexuality to the expression of violence and hostility instead of love. A woman has the sacred right to control her own body, as does a man. No one has the right to force or coerce another.

Life is valued in Witchcraft, and it is approached with an attitude of joy and wonder, as well as a sense of humor. Life is seen as the gift of the Goddess. If suffering exists, it is not our task to reconcile ourselves to it, but to work for change.

Magic, the art of sensing and shaping the subtle, unseen forces that flow through the world, of awakening deeper levels of consciousness beyond the rational, is an element common to all traditions of Witchcraft. Craft rituals are magical rites: They stimulate an awareness of the hidden side of reality, and awaken long-forgotten powers of the human mind. . . .

Mother-Goddess is reawakening, and we can begin to recover our primal birthright, the sheer, intoxicating joy of being alive. We can open new eyes and see that there is nothing to be saved *from*, no struggle of life *against* the universe, no God outside the world to be feared and obeyed; only the Goddess, the Mother, the turning spiral that whirls us in and out of existence, whose winking eye is the pulse of being – birth, death, rebirth – whose laughter bubbles and courses through all things and who is found only through love: love of trees, of stones, of sky and clouds, of scented blossoms and thundering waves; of all that runs and flies and swims and crawls on her face; through love of ourselves; life-dissolving world-creating orgasmic love of each other; each of us unique and natural as a snowflake, each of us our own star, her Child, her lover, her beloved, her Self.

The Spiral Dance, New York: Harper & Row (1979), Introduction.

Syed Hassein Nasr

Islam and the Earth

When we turn to Islam we find a religious tradition more akin to Christianity in its theological formulations yet possessing in its heart a gnosis or *sapientia* similar to the metaphysical doctrines of other Oriental traditions. In this, as in many other domains, Islam is the 'middle people', the *ummah wasaṭah* to which the Quran refers, in both a geographical and metaphysical sense. For this reason the intellectual structure of Islam and its cosmological doctrines and sciences of nature can be of the greatest aid in awakening certain dormant possibilities within Christianity.[34]

One finds in Islam an elaborate hierarchy of knowledge integrated by the principle of unity (*al-tawḥīd*) which runs as an axis through every mode of knowledge and also of being. There are juridical, social and theological sciences; and there are gnostic and metaphysical ones all derived in their principles from the source of the revelation which is the Quran. Then there have developed within Islamic civilization, elaborate philosophical, natural and mathematical sciences which became integrated into the Islamic view and were totally Muslimized. On each level of knowledge nature is seen in a particular light. For the jurists and theologians (*mutakallimūn*) it is the background for human action. For the philosopher and scientist it is a domain to be analyzed and understood. On the metaphysical and gnostic level it is the object of contemplation and the mirror reflecting supra-sensible realities.[35]

Moreover, there has been throughout Islamic history an intimate connection between gnosis, or the metaphysical dimension of the tradition, and the study of nature as we also find it in Chinese Taoism. So many of the Muslim scientists like Avicenna, Quṭb al-Dīn Shīrāzī and Bahā' al-Din 'Āmilī were either practising Sufis or were intellectually attached to the illuminationist–gnostic schools. In Islam as in China observation of nature and even experimentation stood for the most part on the side of the gnostic and mystical element of the tradition while logic and rationalistic thought usually remained aloof from the actual observation of nature. There never occurred the alignment found in seventeenth-century science, namely a wedding of rationalism and empiricism which however was now totally divorced from the one experiment that was central for the men of old, namely experiment with oneself through a spiritual discipline.[36]

In Islam the inseparable link between man and nature, and also between the sciences of nature and religion, is to be found in the Quran itself, the Divine Book which is the Logos or the Word of God. As such it is both the source of the revelation which is the basis of religion and that macrocosmic revelation which is the Universe. It is both the recorded Quran (*al-Qur'ān al-tadwīnī*) and the 'Quran of creation' (*al-Qur'ān al-takwīnī*) which contains the 'ideas' or archetypes of all things. That is why the term used to signify the verses of the Quran or *āyah* also means events occurring within the souls of men and phenomena in the world of nature.[37]

Revelation to men is inseparable from the cosmic revelation which is also a book of

God. Yet the intimate knowledge of nature depends upon the knowledge of the inner meaning of the sacred text or hermeneutic interpretation (*ta'wīl*).[38] The key to the inner meaning of things lies in *ta'wīl*, in penetrating from the outward (*zāhir*) to the inward (*Bātin*) meaning of the Quran, a process which is the very opposite of the higher criticism of today. The search for the roots of knowledge in the esoteric meaning of a sacred text is also found in Philo and certain medieval Christian authors such as Hugo of St Victor and Joachim of Flora. Outside the mainstream of Christian orthodoxy it is found after the Renaissance in such writers as Swedenborg. It is precisely this tradition, however, that comes to an end in the West with the obliteration of metaphysical doctrines leaving the sacred text opaque and unable to answer the questions posed by the natural sciences. Left only with the external meaning of the Holy Scripture later Christian theologians could find no other refuge than a fundamentalism whose pathetic flight before nineteenth century science is still fresh in the memory.

By refusing to separate man and nature completely, Islam has preserved an integral view of the Universe and sees in the arteries of the cosmic and natural order the flow of divine grace or *barakah*. Man seeks the transcendent and the supernatural, but not against the background of a profane nature that is opposed to grace and the supernatural. From the bosom of nature man seeks to transcend nature and nature herself can be an aid in this process provided man can learn to contemplate it, not as an independent domain of reality but as a mirror reflecting a higher reality, a vast panorama of symbols which speak to man and have meaning for him.[39]

The purpose of man's appearance in this world is, according to Islam, in order to gain total knowledge of things, to become the Universal Man (*al-insān al-kāmil*), the mirror reflecting all the Divine Names and Qualities.[40] Before his fall man was in the Edenic state, the Primordial Man (*al-insān al-qadīm*); after his fall he lost this state, but by virtue of finding himself as the central being in a Universe which he can know completely, he can surpass his state before the fall to become the Universal Man. Therefore, if he takes advantage of the opportunity life has afforded him, with the help of the cosmos he can leave it with more than he had before his fall.

The purpose and aim of creation is in fact for God to come 'to know' Himself through His perfect instrument of knowledge that is the Universal Man. Man therefore occupies a particular position in this world. He is at the axis and centre of the cosmic *milieu* at once the master and custodian of nature. By being taught the names of all things he gains domination over them, but he is given this power only because he is the vicegerent (*khalīfah*) of God on earth and the instrument of His Will. Man is given the right to dominate over nature only by virtue of his theomorphic make-up, not as a rebel against heaven.

In fact man is the channel of grace for nature; through his active participation in the spiritual world he casts light into the world of nature. He is the mouth through which nature breathes and lives. Because of the intimate connection between man and nature, the inner state of man is reflected in the external order.[41] Were there to be no more contemplatives and saints, nature would become deprived of the light that illuminates it and the air which keeps it alive. It explains why, when man's inner being has turned to darkness and chaos, nature is also turned from harmony and beauty to disequilibrium and disorder.[42] Man sees in nature what he is himself and penetrates into the inner meaning of nature only on the condition of being able to delve into the inner depths of his own being and to cease to lie merely on the periphery of his being. Men who live only on the surface of their being can study nature as something to be manipulated and dominated. But only he who has turned toward the inward dimension of his being can see nature as a symbol, as a transparent reality and come to know and understand it in the real sense.

Notes

37 In fact the Quran asserts, 'We shall show them our portents upon the horizons and within themselves, until it be manifest unto them that it is the Truth'. (XLI; 53) (Pickthall translation); see Nasr, *An Introduction to Islamic Cosmological Doctrines*, p. 6.

38 See H. Corbin (with the collaboration of S. H. Nasr and O. Yahya), *Histoire de la philosophie islamique*, Paris, 1964, pp. 13–30; and H. Corbin, 'L'intériorisation du sense en herméneutique soufie iranienne', *Eranos Jabrhuch*, XXVI, Zurich, 1958. See also S. H. Nasr, *Ideals and Realities of Islam*, London, 1966, chapter II.

39 'Nor is there anything which is more than a shadow. Indeed, if a world did not cast down shadows from above, the worlds below it would at once vanish altogether, since each world in creation is no more than a tissue of shadows entirely dependent on the archetypes in the world above. Thus the foremost and truest fact about any form is that it is a symbol, so that when contemplating something in order to be reminded of its higher realities the traveller is considering that thing in its universal aspect which alone explains its existence.' Abu Bakr Siraj Ed-Din, *The Book of Certainty*, London, 1952, p. 50.

40 On this capital doctrine see al-Jili, *De l'homme universel* (trans. T. Burckhardt), Lyon, 1953; and T. Burckhardt, *An Introduction to Sufi Doctrine* (trans. D.M. Matheson), Lahore, 1959.

41 'In considering what the religions teach, it is essential to remember that the outside world is as a reflection of the soul of man . . .' *The Book of Certainty*, p. 32. 'The state of the outer world does not merely correspond to the general state of men's souls; it also in a sense depends on that state, since man himself is the pontiff of the outer world. Thus the corruption of man must necessarily affect the whole, . . .' *Ibid.*, p. 33.

42 A traditional Muslim would see in the bleakness and ugliness of modern industrial society and the ambience it creates an outward reflection of the darkness within the souls of men who have created this order and who live in it.

The Encounter of Man and Nature, London: Allen & Unwin (1968), 94–6.

CHAPTER 16

Literary roots

Sappho, Come hither to me from Crete
R. Graves, The White Goddess
H. D. Thoreau, Walden
N. Carpenter, Guild Socialism

Introduction

> Cynicism and disaffection are only half the story of the counter-culture – the first half, the initial act of outraged and desperate rejection. Beyond the 'no' there has been a 'yes' emerging, the affirmation of a vision of life that belongs to a continuing, subterranean current of unorthodox thought and art which reaches back in Western society at least to the Romantic movement, if not into far older traditions ... My interest and conviction have been with those disenting elements who have climbed aboard William Blake's chariot of fire to join his 'mental fight' against the total secularization and scientization of modern culture.
>
> (Roszak 1979: 19)

Literary essays, novels, poetry and short stories have all been used to transmit ecological and holistic concepts. Literature reaches further than a manifesto, philosophical tract or scientific report, and deeper. We have already noted the Green sensibilities of the Romantic poets, especially Blake, Goethe, Shelley and Wordsworth. The list of Green novelists may be said to include Aldous Huxley and D. H. Lawrence. Tolstoy, the anarchist, vegetarian advocate of non-violence, voiced his love of the natural world even in such epic works as *War and Peace*. Yet even the 'manifestoes', polemical works written without pretensions to literary merit, include much that can be described as literature. We may also include the works of Carlyle, Ruskin and Morris, examples of which may be found elsewhere in this volume.

Poetry, it can be argued, is one of the most important vehicles of Green Idealism. Robert Graves, author of *I, Claudius* and himself a poet, contended that poetry's origins could be found in the worship of the moon-goddess or muse. He believed that poetic myths containing themes from the Old Stone Age were eroded by the arrival of 'invaders from Central Asia', who ended the matriarchal and implicitly environmentally balanced societies of Europe. In his opinion these invaders degraded nature, language and poetry, introducing a new order of logic, looking to male gods and creating in turn a stratified society. Such a new order was, for him, exemplified by classical Greek civilization. The holy literature of pre-classical Europe survived, for Graves, in the form of the Welsh epics like *The Song of Amergin*.

Sappho (610–580 B.C.), a woman who wrote beautiful poems of which only fragments survive, produced verse often based around nature themes, despite the rule of Apollo. Poems revering nature, if not the Moon Goddess or Gaia, can be found throughout the millenia that followed the fall of Minoan Crete. As well as Wordsworth and Gerald Manley

Hopkins we may also mention the Japanese Haiku. Allen Ginsberg, enlightened by the voice of Blake in his Harlem tenement, wrote *Sunflower Sutra*, a work which probes beyond appearances to find the shining sunflower beneath the grime and reveal the vitality of human life and nature behind superficial industrial adversity.

The utopian novel has been used to project a political or philosophical message. Such novels, portraying a perfect society of social harmony, justice and environmental balance, are discussed in our final chapter. D. H. Lawrence, among others, debated his political beliefs in novels like *Kangaroo*, while in *Unto this Last* and *The Stones of Venice* Ruskin produced critical essays that count as literature. Henry David Thoreau's account of his life in the woods of Walden counts as a Green classic. The journal contains observations of elementary ecological science in the tradition of Gilbert White and other philosophical, political and economic conclusions. Here in a number of passages we find Thoreau discussing the merits of vegetarianism, the need to live a simple life, the quiet desperation of the mass of humanity. One suspects that if he had attempted to write a didactic utopian work the effect would have been far less interesting.

Carpenter in his study of *Guild Socialism*, a movement that sprang up after the death of William Morris, for a while continued his campaign for the creation of work that is environmental, creative and controlled by the labourer. He traces the literary impetus behind such a vision. His account of Carlyle, Ruskin and others provided a pedigree for ideas that have informed the modern Greek movement, just as much as the Guild Socialism of the early twentieth century.

Sappho

Come hither to me from Crete

Come hither to me from Crete,
to this holy temple,
where is your graceful grove of apple-trees,
and altars fragrant with incense;

A place where cold water babbles through apple-branches,
shadowed by roses,
from whose quivering leaves comes sleep;

With a meadow blossoming with flowers,
where horses pasture,
and there a breeze blows sweetly;

Come, Cyprian goddess,
softly pour nectar,
into our cups for feasting.

Sappho (seventh century B.C.), a fragment of a poem found on a piece of broken pottery; adapted from translations by Bowra (1961: 196–7), Kirkwood (1974: 114–15) and Page (1955: 34–5).

Robert Graves

The White Goddess

My thesis is that the language of poetic myth anciently current in the Mediterranean and Northern Europe was a magical language bound up with popular religious ceremonies in honour of the Moon-goddess, or Muse, some of them dating from the Old Stone Age, and that this remains the language of true poetry – 'true' in the nostalgic sense of 'the unimprovable original, not a synthetic substitute'. The language was tampered with in late Minoan times when invaders from Central Asia began to substitute patrilinear for matrilinear institutions and remodel or falsify the myths to justify social changes. Then came the early Greek philosophers, who were strongly opposed to magical poetry as threatening their new religion of logic, and under their influence a rational poetical language (now called the Classical) was elaborated in honour of their patron Apollo and imposed on the world as the last word in spiritual illumination: a view that has prevailed practically ever since in European schools and universities, where myths are now studied only as quaint relics of the nursery age of mankind.

One of the most uncompromising rejections of early Greek mythology was made by Socrates. Myths frightened or offended him; he preferred to turn his back on them and discipline his mind to think scientifically: 'to investigate the reason of the being of every-thing – of everything as it is, not as it appears, and to reject all opinions of which no account can be given'. . . .

I deduce from the petulant tone of his phrase 'vulgar cleverness' that he had spent a long time worrying about the Chimaera, the horse-centaurs and the rest, but that the 'reasons of their being' had eluded him because he was no poet and mistrusted poets, and because, as he admitted to Phaedrus, he was a confirmed townsman who seldom visited the countryside: 'fields and trees will not teach anything, but men do'. The study of mythology, as I shall show, is based squarely on tree-lore and seasonal observation of life in the fields. . . .

But even after Alexander the Great had cut the Gordian Knot – an act of far greater moral significance than is generally realized – the ancient language survived purely enough in the secret Mystery-cults of Eleusis, Corinth, Samothrace and elsewhere; and when these were suppressed by the early Christian Emperors it was still taught in the poetic colleges of Ireland and Wales, and in the witch-covens of Western Europe. As a popular religious tradition it all but flickered out at the close of the seventeenth century; and though poetry of a magical quality is still occasionally written, even in industrialized Europe, this always results from an inspired, almost pathological, reversion to the original language – a wild Pentecostal 'speaking with tongues' – rather than from a conscientious study of its grammar and vocabulary.

English poetic education should, really, begin not with the *Canterbury Tales*, not with the *Odyssey*, not even with Genesis, but with the *Song of Amergin*, an ancient Celtic calendar-alphabet, found in several purposely garbled Irish and Welsh variants, which briefly summarizes the prime poetic myth. I have tentatively restored the text as follows:

I am a stag: *of seven times,*
I am a flood: *across a plain,*
I am a wind: *on a deep lake,*
I am a tear: *the Sun lets fall,*
I am a hawk: *above the cliff,*
I am a thorn: *beneath the nail,*
I am a wonder: *among flowers,*
I am a wizard: *who but I*
Sets the cool head aflame with smoke?

I am a spear: *that roars for blood,*
I am a salmon: *in a pool,*
I am a lure: *from paradise,*
I am a hill: *where poets walk,*
I am a boar: *ruthless and red,*
I am a breaker: *threatening doom,*
I am a tide: *that drags to death,*
I am an infant: *who but I*
Peeps from the unhewn dolmen arch?

I am the womb: *of every holt,*
I am the blaze: *on every hill,*
I am the queen: *of every hive,*
I am the shield: *for every head,*
I am the tomb: *of every hope.*

It is unfortunate that, despite the strong mythical element in Christianity, 'mythical' has come to mean 'fanciful, absurd, unhistorical'; for fancy played a negligible part in the development of the Greek, Latin and Palestinian myths, or of the Celtic myths until the Norman–French *trovères* worked them up into irresponsible romances of chivalry. They are all grave records of ancient religious customs or events, and reliable enough as history once their language is understood and allowance has been made for errors in transcription, misunderstanding of obsolete ritual, and deliberate changes introduced for moral or political reasons. . . .

'What is the use of poetry nowadays?' is a question not the less poignant for being defiantly asked by so many stupid people or apologetically answered by so many silly people. The function of poetry is religious invocation of the Muse; its use is the experience of mixed exaltation and horror that her presence excites. But nowadays? Function and use remain the same; only the application has changed. This was once a warning to man that he must keep in harmony with the family of living creatures among which he was born, by obedience to the wishes of the lady of the house; it is now a reminder that he has disregarded the warning, turned the house upside down by capricious experiments in philosophy, science and industry, and brought ruin on himself and his family. 'Nowadays' is a civilization in which the prime emblems of poetry are dishonoured. In which serpent, lion and eagle belong to the circus-tent; ox, salmon and boar to the cannery; racehorse and greyhound to the betting ring; and the sacred grove to the sawmill. In which the Moon is despised as a burned-out satellite of the Earth and woman reckoned as 'auxiliary State personnel'. In which money will buy almost anything but truth, and almost anyone but the truth-possessed poet.

Call me, if you like, the fox who has lost his brush; I am nobody's servant and have chosen to live on the outskirts of a Majorcan mountain village, Catholic but anti-ecclesiastical, where life is still ruled by the old agricultural cycle. Without my brush, namely my contact with urban civilization, all that I write must read perversely and irrelevantly to such of you as are still geared to the industrial machine, whether directly as workers, managers, traders or advertisers or indirectly as civil servants, publishers, journalists, schoolmasters or employees of a radio corporation. If you are poets, you will realize that acceptance of my historical thesis commits you to a confession of disloyalty which you will be loth to make; you chose your jobs because they promise to provide you with a steady income and leisure to render the Goddess whom you adore valuable part-time service. Who am I, you will ask, to warn you that she demands either whole-time service or none at all? And do I suggest that you should resign your jobs and, for want of sufficient capital to set up as smallholders, turn romantic shepherds – as Don Quixote did after his failure to come to terms with the modern world – in remote unmechanized farms? No, my brushlessness debars me from offering any practical suggestion. I dare attempt only a historical statement of the problem; how you come to terms with the Goddess is no concern of mine. I do not even know that you are serious in your poetic profession.

The White Goddess (1948), London: Faber (1961), 9–15.

Henry David Thoreau

Walden

The mass of men lead lives of quiet desperation. What is called resignation is confirmed desperation. From the desperate city you go into the desperate country, and have to console yourself with the bravery of minks and muskrats. A stereotyped but unconscious despair is concealed even under what are called the games and amusements of mankind. There is no play in them, for this comes after work. But it is a characteristic of wisdom not to do desperate things. . . .

The greater part of what my neighbours call good I believe in my soul to be bad, and if I repent of anything, it is very likely to be my good behaviour. What demon possessed me that I behaved so well? You may say the wisest thing you can, old man – you who have lived seventy years, not without honour of a kind – I hear an irresistible voice which invites me away from all that. One generation abandons the enterprises of another like stranded vessels.

I think that we may safely trust a good deal more than we do. We may waive just so much care of ourselves as we honestly bestow elsewhere. Nature is as well adapted to our weakness as to our strength.... Let us consider for a moment what most of the trouble and anxiety which I have referred to is about, and how much it is necessary that we be troubled, or at least, careful. It would be some advantage to live a primitive and frontier life, though in the midst of an outwards civilisation, if only to learn what are the gross necessaries of life and what methods have been taken to obtain them; or even to look over the old day-books of the merchants, to see what it was that the men most commonly bought at the stores, what they stored, that is, what are the grossest groceries. For the improvements of ages have had but little influence on the essential laws of man's existence: as our skeletons, probably, are not to be distinguished from those of our ancestors.

By the words, *necessary of life*, I mean whatever, of all that man obtains by his own exertions, has been from the first, or from long use has become, so important to human life that few, if any, whether from savageness, or poverty, or philosophy, ever attempt to do without it.... Most of the luxuries, and many of the so-called comforts of life, are not only not indispensable, but positive hindrances to the elevation of mankind. With respect to luxuries and comforts, the wisest have ever lived a more simple and meagre life than the poor. The ancient philosophers, Chinese, Hindoo, Persian, and Greek, were a class than which none has been poorer in outward riches, none so rich in inward....

Every morning was a cheerful invitation to make my life of equal simplicity, and I may say innocence, with Nature herself. I have been as sincere a worshipper of Aurora as the Greeks. I got up early and bathed in the pond: that was a religious exercise, and one of the best things which I did. They say that characters were engraven on the bathing tub of king Tching-thang to this effect: 'Renew thyself completely each day; do it again, and again, and forever again.' I can understand that. Morning brings back the heroic ages. I was as much affected by the faint hum of a mosquito making its invisible and unimaginable tour through my apartment at earliest dawn, when I was sitting with door and windows open, as I could be by any trumpet that ever sang of fame. It was Homer's requiem; itself an Iliad and Odyssey in the air, singing its own wrath and wanderings. There was something cosmical about it; a standing advertisement, till forbidden, of the everlasting vigour and fertility of the world. The morning, which is the most memorable season of the day, is the awakening hour. Then there is least somnolence in us; and for an hour, at least, some part of us awakes which slumbers all the rest of the day and night. Little is to be expected of that day, if it can be called a day, to which we are not awakened by our Genius, but by the mechanical nudgings of some servitor, are not awakened by our newly acquired force and aspirations from within, accompanied by the undulations of celestial music, instead of factory bells, and a fragrance filling the air – to a higher life than we fell asleep from; and thus the darkness bear its fruit, and prove itself to be good, no less than the light. That man who does not believe that each day contains an earlier, more sacred, and auroral hour than he has yet profaned, has despaired of life, and is pursuing a descending and darkening way....

I went to the woods because I wished to live deliberately, to front only the essential facts of life, and see if I could not learn what it had to teach, and not, when I came to die, discover that I had not lived. I did not wish to live what was not life, living is so dear; nor did I wish to practise resignation, unless it was quite necessary. I wanted to live deep and suck out all the marrow of life.... Simplicity, simplicity, simplicity! I say, let your affairs be as two or three, and not a hundred or a thousand; instead of a million count half a dozen, and keep your accounts on your thumbnail. In the midst of this chopping sea of civilised life, such are the clouds and storms and quicksand and thousand-and-one items to be allowed for, that man has to live, if he would not founder and go to the bottom and not make his port at all, by dead reckoning, and he must be a great calculator indeed who succeeds. Simplify, simplify....

I had long felt differently about fowling, and sold my gun before I went to the woods. Not that I am less humane than others, but I did not perceive that my feelings were much affected. I did not pity the fishes nor the worms. This was habit. As for fowling, during the last years that I carried a gun my excuse was that I was studying ornithology and sought only new or rare birds. But I confess that I am now inclined to think that there is a finer way of studying ornithology than this....

Is it not a reproach that man is a carnivorous animal? True, he can and does live, in great measure, by preying on other animals; but this is a miserable way, – as any one who will go to snaring rabbits, or slaughtering lambs, may learn, – and he will be regarded as a benefactor of his race who shall teach man to confine himself to a more innocent and wholesome diet. Whatever my own practice may be, I have no doubt that it is a part of the destiny of the human race, in its gradual improvement, to leave off eating animals, as surely as the savage tribes have left off eating each other when they came into contact with the more civilised.

Walden (1854), London: Dent (1965), 5, 8–11, 76–7, 186, 190.

N. Carpenter

Guild Socialism

Born at the close of the eighteenth century, Carlyle, by his life and observations, gained an intimate knowledge of the sufferings of the English and Scotch factory operatives in the early nineteenth century.[1] He became acquainted, too, with their turbulent struggles for freedom, and acquired not a little of their spirit;[2] also, in common with other serious-minded men of his day, he gave ear to the proposals put forward both by Whig 'radicals', and by Tories.

His reaction was noteworthy. He repudiated what he considered the godless and hypocritical spirit of his age. As will be shown later, he turned back, instead, to the ideals of another period. But he also kept his eyes clearly on the present, and launched against the theory and practice of 'free competition' a ringing denunciation. His social ideas were expressed especially in such works as *Signs of the Times*, published in 1829; *Chartism*; published in 1839, and *Past and Present*, written in 1843.

While roundly denouncing the excesses of the Chartists, Carlyle maintained that their agitation could not be permanently 'put down' until the injustices and sufferings, which had driven 'fierce and mad' the 'great, dumb, deep-buried class' of common people, had been removed.

He saw a root cause of their sufferings in bad government. In industry, this condition arose from the 'cash-nexus', that is, the loosing of all ties between master and workmen save merely the wage-relation, and the consequent refusal by the master to have any concern whatever for the welfare of his employees, in his own mines and mills, or in their homes. In politics, Carlyle found the same desertion of the masses by those responsible for them, only here the principle at fault was the doctrine of *laissez-faire*. He inveighed passionately also, against the economic theories justifying such a policy, especially that school of Malthusians which, in his day, had become so obsessed by the idea of 'over-population', that one of their number solemnly proposed the painless death of all children of a poor family numbering more than three.[3]

Carlyle found another cause of the workingmen's woes in the general absence of a proper purpose of life. Instead of the 'gospel of work' in which he believed,[4] he saw men dominated by a false utilitarian philosophy, and – particularly in the case of the rich – by 'Mammonism'.[5]

All of this Carlyle wrote with a vehement fervor and a vivid directness which rather dazed and shocked most of his readers, so that his direct influence was probably little. But to the younger generation of writers and reformers, he appeared as a prophet, and his authority grew, rather than lessened, as the years passed. In his own lifetime the Christian Socialists, the novelists, and even the Chartist poets acknowledged their debt to him.[6] In a later generation he exercised an influence over such men as Ruskin, Morris, and still later, the founders of the Guild Movement, the importance of which can only be conjectured.

Ruskin may be said directly to have carried on the work of Carlyle.[7] Like Carlyle, Ruskin compared Medieval England with that of his own times – to the disadvantage of the latter. Ruskin criticized current economic conditions from a viewpoint, however, distinct from Carlyle's, though he reached similar conclusions. He attacked the system on its theoretical basis: first, the political economy which sought to justify and explain it, and, second, the whole principle of large-scale production.

He objected to the ruling school of economics because it seemed to him to depend on false values, particularly the substitution of material wealth for social wealth, or welfare, the emphasis on self-interest, and the ignoring of the human side of labor – a feature of his doctrine which unites him with the economists of revolt already discussed. Ruskin attempted to build up, in opposition to the 'Manchester' political economy, a system of his own, in which *human values* were given first place....[8]

Matthew Arnold had much in common with Carlyle and Ruskin. He expressed an aversion, not so much to the workings of the English factory system, as to the motives of those directing it. He found, in the materialism of the age, and in its exaltation of machinery – whether in industry or in politics – a spirit of 'Philistinism', and contrasted it with the spirit of 'Hellenism', for which he pleaded.[9]

Notes

1   Craig, *The Making of Carlyle* (New York, 1909), pp. 116, 284–6.
2   *Ibid.*, pp. 284, 285.
3   Postgate, *Revolution*, p. 122, footnote.
4   *Vide Sartor Resartus* (Everyman's Edition), pp. 148–9.
5   Carlyle, *Chartism* (London, 1839), and *Past and Present, passim. Cf.* especially the account of

'Plugson of Undershot' in *Past and Present*, Book III.

6   Woodworth, *Christian Socialism*, p. 10; Kauffman, *Charles Kingsley* (London, 1892), pp. 181 ff; Beer, *British Socialism*, Vol. II, p. 105. Dickens' *Hard times* was dedicated to Carlyle, and shows his influence in matters of phraseology. For an account of the influence of Carlyle on Disraeli, cf. Sichel, *Disraeli* (London, 1904), pp. 85–92.

7   *Cf.* his apostrophe to Carlyle, *Munera Pulveris*, sec. 159.

8   Ruskin, *Munera Pulveris* (London, 1871), especially secs. 11–14, 59; and *Unto this Last* (London, 1862), *passim*.

9   Arnold, *Culture and Anarchy* (London, 1869), *passim*.

Guild Socialism, New York: Appleton (1922), 28–30, 31.

CHAPTER 17

Green revolutionaries

G. Winstanley, A declaration from the poor oppressed people of England

H. Stretton, Capitalism, socialism and nature

R. Blatchford, Merrie England

L. G. Wilkinson, Women in freedom

Introduction

> What we call man's power over nature turns out to be a power exercised by some men over other men with Nature as its instrument.
>
> (C.S. Lewis 1943: 28)

Many radical political movements, as well as stressing humanist goals of greater social justice and economic quality, have on occasions integrated Green themes such as ecological awareness, decentralization and animal liberation. Greens have had to look, too, beyond the symptoms of environmental destruction to deeper causes: economic systems that demand the exploitation of nature and of the work force, undemocratic centralized structures that rob individuals of control over their surroundings, and a militarism that damages humanity and nature. Constitutional Green parties orientated around the goal of achieving parliamentary representation have been a response to this deepening and widening of the Green outlook. Another has been that of direct action and grass-root struggles for a new social order.

The modern Green movement shares many revolutionary goals with older peasant ideologies. From the uprisings of medieval peasants to the creation of a Green International of European peasant parties, many have argued that individuals should have close contact with the land (Canovan 1977: 88). Perhaps the most sophisticated and radical expression of such politicized peasantry can be found with our first extract. During the turmoil of the English civil war of the seventeenth century the Diggers, a radical movement led by Gerrard Winstanley (1609–76?), sought to cultivate empty land that had been abandoned or left fallow by large landowners. The Diggers were ruralists who believed that all should work on the land and that agriculture provided the potential of good work for all. Being communistic, they believed in a co-operative society. They were also decentralists and pacifists. Above all they practised non-violent direct action, occupying and farming derelict land on St George's Hill in Surrey. It is difficult, despite these similarities, to label them as Greens. Winstanley did, after all, see the earth as 'a common storehouse for all'. Although like the Native American he opposed the idea that land might 'be bought or sold', he still essentially saw the natural environment as a gift from God to humanity to use as humanity wished.

Our second extract, from Stretton's study *Capitalism, Socialism and Environment*, written imaginatively from the perspective of a future conservative ecological order, describes an age when 'the rich have learned to welcome the death of the poor'. This is a convenient process that provides them with scarce resources in a finite world. In Stretton's fable the one remaining enclave of freedom has produced a school of scholars who examine the process whereby earlier environmentally conscious aristocrats had cleared Britain of the poor, expelling them to the cities or exiling them to the colonies. With this long and bloody expropriation 'the most beautiful landscape in English history was made ready for the Romantic poets'. The diverse and cloudy concept of Romanticism can be used, as we have already noted, to provide Green forefathers and mothers, concerned to defend both the landscape and its people. Blake, Shelley, Goethe, Morris and D. H. Lawrence were all in a sense Romantics, adhering to essential Green concepts. The conservatives Carlyle and Ruskin, who believed in recreating the feudal order, albeit with a little more equality than existed in medieval Europe, ironically inspired both Greens and early British socialists. Shelley and Blake were certainly radicals who combined their Green with deeper shades of Red. Shelley, as we have noted, was influenced by his father-in-law, William Godwin (1756–1836), whose *Political Justice* put the case for a decentralized anarchist society, composed of 'a small community or parish inhabited by a cultured peasantry, each family subsisting upon smallholdings of equal size and attending with equal devotion to the husbandry of the soil and the mind' (Willey 1949: 222).

A new generation of radicals, examined by Gould in his *Early Green Politics*, shared virtually all the values of the modern movement. Members of this current, which flourished between 1880 and 1900, supply our final extracts. Robert Blatchford's *Merrie England* is a tract that attacked capitalism for separating humanity from nature and producing goods that were artificial and unnecessary. The book sold a million copies between 1894 and 1895, while Blatchford's journal the *Clarion* had the widest circulation of any socialist newspaper of the time. Blatchford popularized the values of Edward Carpenter and William Morris amongst British working men and women at the end of the nineteenth century. Clarion cycling clubs drew tens of thousands of supporters on to the roads during the warm summer days of the Edwardian era. In the 1930s former members of the Clarion clubs took part in the campaign to gain access to private land for walkers and climbers. This in turn echoed the demands of the Diggers, who affirmed that the land should be held in common for the use and care of all people (Stephenson 1989). Our final extract comes from Lily Gair Wilkinson, who in 1900 combined socialism and ecological concern with the demand for women's liberation.

Gerrard Winstanley

A declaration from the poor oppressed people of England

We whose names are subscribed, do in the name of all the poor oppressed people in *England*, declare unto you, that call your selves Lords of Manors, and Lords of the Land, That in regard the King of Righteousness, our Maker, hath inlightened our hearts so far, as to see, That the earth was not made purposely for you, to be Lords of it, and we to be your Slaves, Servants, and Beggers; but it was made to be a common Livelihood to all, without respect of persons: And that your buying and selling of Land, and the Fruits of it, one to another, is *The cursed thing*, and was brought in by War; which hath, and still does establish murder, and theft, in the hands of some branches of Mankinde over others, which is the greatest outward burden, and unrighteous power, that the Creation groans under: For the power of inclosing Land, and owning Propriety, was brought into Creation by your Ancestors by the Sword; which first did murther their fellow Creatures, Men, and after plunder or steal away their Land, and left this Land successively to you, their Children. And therefore, though you did not kill or theeve, yet you hold that cursed thing in your hand, by the power of the Sword; and so you justifie the wicked deeds of your Fathers; and that sin of your Fathers, shall be visited upon the Head of you, and your Children, to the third and fourth Generation, and longer too, tell your bloody and theeving power be rooted out of the Land.

And further, in regard the King of Righteousness hath made us sensible of our burthens, and the cryes and groanings of our hearts are come before him: We take it as a testimony of love from him, that our hearts begin to be freed from slavish fear of men, such as you are; and that we finde Resolutions in us, grounded upon the inward law of Love, one towards another, To Dig and Plough up the Commons, and waste Lands through *England*, and that our conversation shall be so unblameable, That your Laws shall not reach to oppress us any longer, unless you by your Laws will shed the innocent blood that runs in our veins.

For though you and your Ancestors got your Propriety by murther and theft, and you keep it by the same power from us, that have an equal right to the Land with you, by the righteous Law of Creation, yet we shall have no occasion of quarreling (as you do) about that disturbing devil, called *Particular Propriety*: For the Earth, with all her Fruits of Corn, Cattle, and such like, was made to be a common Storehouse of Livelihood to all Mankinde, friend and foe, without exception.

And to prevent all your scrupulous Objections, know this, That we must neither buy nor sell; Money must not any longer (after our work of the Earths community is advanced) be the great god, that hedges in some, and hedges out others; for Money is but part of the Earth: And surely, the Righteous Creator, who is King, did never ordain, That unless some of Mankinde, to bring that Mineral (Silver and Gold) in their hands, to others of their own kinde, that they should neither be fed, nor be clothed; no surely, For this was the project of Tyrant-flesh (which Land-lords are branches of) to set his Image upon Money. And they make this unrighteous Law, That none should buy or sell, eat, or be clothed, or have any comfortable Livelihood among men, unless they did bring his Image stamped upon Gold or Silver in their hands.

And whereas the Scriptures speak, That the mark of the Beast is 666, the number of a man; and that those that do not bring that mark in their hands, or in their foreheads, they should neither buy nor sell, *Revel*. 13. 16. And seeing the numbering Letters round about the English money make 666[1], which is the number of that Kingly Power and Glory (called a *Man*), And seeing the age of the Creation is now come to the Image of the Beast, or Half day, And seeing 666 is his mark, we expect this to be the last Tyrannical power that shall raign; and that people shall live freely in the enjoyment of the Earth, without bringing the mark of the Beast in their hands, or in their promise; and that they shall buy Wine and Milk, without Money, or without price, as *Isaiah* speaks.

For after our work of the Earthly community is advanced, we must make use of Gold and Silver, as we do of other mettals, but not to buy and sell withal; for buying and selling is the great cheat, that robs and steals the Earth one from another: It is that which makes some Lords, others Beggers, some Rulers, others to be ruled; and makes great Murderers and Theeves to be imprisoners, and hangers of little ones, or of sincere-hearted men.

And while we are made to labor the Earth together, with one consent and willing minde; and while we are made free, that every one, friend and foe, shall enjoy the benefit of their Creation, that is, To have food and rayment from the Earth, their Mother; and every one subject to give accompt of this thoughts, words and actions to none, but to the one onely righteous Judg, and Prince of Peace, the Spirit of Righteousness that dwells, and that is now rising up to rule in every Creature, and in the whole Globe. We say, while we are made to hinder no man of his Priviledges given him in his Creation, equal to one, as to another; what Law then can you make, to take hold upon us, but Laws of Oppression and Tyranny, that shall enslave or spill the blood of the Innocent? And so your Selves, your Judges, Lawyers, and Justices, shall be found to be the greatest Transgressors, in and over Mankinde.

But to draw neerer to declare our meaning, what we would have, and what we shall endevor to the uttermost to obtain, as moderate and righteous Reason directs us; seeing we are made to see our Priviledges, given us in our Creation, which have hitherto been denied to us, and our Fathers, since the power of the Sword began to rule, And the secrets of the Creation have been locked up under the traditional, Parrat-like speaking, from the Universities, and Colledges for Scholars, And since the power of the murdering, and theeving Sword, formerly, as well as now of late yeers, hath set up a Government, and maintains that Government; for what are prisons, and putting others to death, but the power of the Sword; to enforce people to that Government which was got by Conquest and Sword, and cannot stand of it self, but by the same murdering power? That Government that is got over people by the Sword, and kept by the Sword, is not set up by the King of Righteousness to be his Law, but by Coveteousness . . .

Note

1   I can explain this only on the supposition that the letter M was overlooked. The lettering, with somewhat differing abbreviations on different coins, was as follows: Carolus D. G. Mag. Br. Fr. et Hi. Rex. The significant letters are MDCLXVI = 1666. [Editor's note]

'A declaration from the poor oppressed people of England' (1649), in G. Sabine, ed., *The Works of Gerrard Winstanley*, New York: Russell (1965), 269–71.

H. Stretton

Capitalism, socialism and nature

Some Anglo-Eskimo scholars – the last enclave with an uncensored English-language press – have drawn attention to what they see as the nearest historical parallel to the achievement, namely the general reconstruction of English life and landscape which took place in the century between about 1740 and 1840.

Through most of that time (the Anglo-Eskimos say) political power belonged to a small class of town and country proprietors. They used it by legal, military, commercial and other means to change the ownership and management of about half of the arable land in England – and from a Red Indian point of view a fair proportion of the land in North and Caribbean America. These transfers, together with the new techniques applied to the consolidated acres, were used to establish radically new scales of domestic and inter-national inequality. Subsistence levels for the masses held steady or even fell; for a century the lion's share of all improvement was creamed off by the proprietors through rent, wage, industrial, tax and military policies of unparalleled profitability. Through much of central England one whole agrarian eco-system was replaced by another. In little more than a generation the new proprietors and their tenants built what is now remembered nostalgically as the 'immemorial' English country landscape of hedged and ditched fields, naturally fertilized rotational farming, protected waters and forests, lonely coasts and highlands, and the stable eco-systems of permanently stocked partridge woods, fox coverts, deer parks, grouse moors and trout and salmon streams.

In the course of the change, many common rights to arable land and pasture, fuel, fish and game were converted to private property by Act of Parliament or other legal process. Of course such transfers of wealth and rights, with new extremes of hardship and inequality, required severe enforcement. At home the death penalty was extended to three times as many offences as formerly. Record numbers crowded the prison hulks and penal colonies. Labor discipline was harsh, at home and overseas. For every white life in England in that century there was an early black death in the separate facility across the Atlantic. But the most beautiful landscape in English history was ready for the romantic poets. Unhappily a good deal of it was soon spoiled again; but it might all have been there still if the steam engine had not started other changes.

The makers of the new environmental revolution and the new environmental landscape have taken better care of that latter sort of danger. They permit very little research or technical change, and they keep what there is of it under watchful political direction. Contraceptive research continues and simple methods have been developed for early abortion and for determining the sex of children at conception. These are the main items in the reformed societies' international aid programs. Poor societies are helped to contracept, to abort, and (less successfully) to implement social preferences for male children. A 'three to one' campaign made some progress in one or two territories where the white profes-sionals in the aid services played tricks with the contraceptive and sex-determinant pills.

The programs were doubly effective while they lasted. As the imbalance developed within the native peasantry and proletariat the female minorities were brothelized (as expected) but also enslaved and overworked to keep up the farming and domestic services traditionally expected of women in poor societies, so they developed abnormal rates of miscarriage, exhaustion and early death. In all cases the males rebelled before long, and the recoil often hit the contraceptive as well as the sex-determinant services.

Meanwhile no aid is given to increase conventional production, consumption or length of life. News of African and Asian famine is received gravely in the reformed white world – but as good news nonetheless. In the interest of humanity, the rich have learned to welcome the death of the poor.

Capitalism, Socialism and Nature, Cambridge: Cambridge University Press (1976), 37–9.

Robert Blatchford

Merrie England

Before we begin this chapter I must ask you to keep in mind the fact that a man's bodily wants are few.

I shall be well outside the mark if I say that a full-grown healthy man can be well fed upon a daily ration of 1lb of bread, 1lb of vegetables, 1lb of meat. Add to this a few groceries, a little fruit, some luxuries, in the shape of wine, beer, and tobacco; a shelter, a bed, some clothing, and a few articles of furniture, and you have all the material things you need.

Remember, also, that when you have got these things you have got all the material things you can use. A millionaire or a monarch could hardly use more, or if he did use more would use them to his hurt and not to his advantage.

You live in Oldham and work in the factory in order to get a living. 'A living' consists of the things above named.

I ask you, as a practical, sensible man, whether it is not possible to get those few simple things with less labour and whether it is not possible to add to them health and the leisure to enjoy life and develop the mind?

The Manchester School will tell you that you are very fortunate to get as much as you do, and that he is a dreamer or a knave who persuades you that you can get more.

The Manchester School is the Commercial School. The supporters of that school will tell you that you cannot prosper, that is to say you cannot 'get a living', without the capitalist, without open competition, and without a great foreign trade.

They will tell you that you would be very foolish to raise your own foodstuffs here in England so long as you can buy them more cheaply from foreign nations. They will tell you that this country is incapable of producing enough food for her present population, and that therefore your very existence depends upon keeping the foreign trade in your hands.

Now, I shall try to prove to you that every one of these statements is untrue. I shall try to satisfy you that:

1. The capitalist is a curse, and not a blessing
2. That competition is wasteful, and cruel, and wrong
3. That no foreign currency can sell us food more cheaply than we can produce it; and
4. That this country is capable of feeding more than treble her population.

We hear a great deal about the value and extent of our foreign trade, and are always being reminded how much we owe to our factory system, and how proud of it we ought to be.

I despise the factory system, and denounce it as a hideous, futile, and false thing. This is one of the reasons why the Manchester School call me a dreamer and a dangerous agitator. I will state my case to you plainly, and ask you for a verdict in accordance with the evidence.

My reasons for attacking the factory system are:

1. Because it is ugly, disagreeable, and mechanical.
2. Because it is injurious to public health.
3. Because it is unnecessary.
4. Because it is a danger to the national existence.

The Manchester School will tell you that the destiny of this country is to become 'The Workshop of the World'.

I say that is not true; and that it would be a thing to deplore if it were true. The idea that this country is to be the 'Workshop of the World' is a wilder dream than any that the wildest socialist ever cherished. But if this country did become the 'Workshop of the World' it would at the same time become the most horrible and the most miserable country the world has ever known.

Let us be practical and look at the facts.

First, as to the question of beauty and pleasantness. You know the factory districts of Lancashire. I ask you, is it not true that they are ugly and dirty, smoky, and disagreeable? Compare the busy towns of Lancashire, of Staffordshire, of Durham, and of South Wales, with the country towns of Surrey, Suffolk and Hants.

In the latter counties you will get pure air, bright skies, clear rivers, clean streets, and beautiful fields, woods, and gardens; you will get cattle and streams, and birds and flowers, and you know that all these things are well worth having, and that none of them can exist side by side with the factory system.

I know that the Manchester School will tell you that this is mere 'sentiment'. But compare their actions with their words.

Do you find the champions of the factory system despising nature, and beauty, and art, and health – except in their speeches and lectures to you?

No. You will find these people living as far as from the factories as they can get; and you will find them spending their long holidays in the most beautiful parts of England, Scotland, Ireland, or the Continent.

The pleasures they enjoy are denied to you. They preach the advantages of the factory system because they reap the benefits while you bear the evils.

To make wealth for themselves they destroy the beauty and the health of your dwelling-places; and then they sit in their suburban villas, or on the hills and terraces of the lovely southern counties, and sneer at the 'sentimentality' of the men who ask you to cherish beauty and to prize health.

Or they point out to you the value of the 'wages' which the factory system brings you, reminding you that you have carpets on your floors, and pianos in your parlours, and a week's holiday at Blackpool once a year.

But how much health or pleasure can you get out of a cheap and vulgar carpet? And what is the use of a piano if you have neither leisure nor means to learn to play it? And why should you prize that one week in the crowded, noisy watering-place, if health and fresh air and the great salt sea are mere sentimental follies?

And let me ask you, is any carpet so beautiful or so pleasant as a carpet of grass and daisies? Is the fifth-rate music you play upon your cheap piano as sweet as the songs of the gushing streams and joyous birds? And does a week at a spoiled and vulgar watering-place repay you for fifty-one weeks' toil and smother in a hideous and stinking town?

As a practical man, would you of your own choice convert a healthy and beautiful country like Surrey into an unhealthy and hideous country like Wigan or Cradley, just for the sake of being able once a year to go to Blackpool, and once a night to listen to a cracked piano?

Now I tell you, my practical friend, that you ought to have, and may have, good music, and good homes, and a fair and healthy country, and more of all the things that make life sweet; that you may have them at less cost of labour than you now pay for the privilege of existing in Oldham; and that you can never have them if England becomes the 'Workshop of the World'.

But the relative beauty and pleasantness of the factory and country districts do not need demonstration. The ugliness of Widnes and Sheffield and the beauty of Dorking and Monsal Dale are not matters of sentiment nor of argument – they are matters of fact. The value of beauty is not a matter of sentiment; it is a fact. You would rather see a squirrel than a sewer rat. You would rather bathe in the Avon than in the Irwell. You would prefer the fragrance of a rose-garden to the stench of a sewage works. You would prefer Bolton Woods to Ancoats slums.

As for those who sneer at beauty, as they spend fortunes on pictures, on architecture, and on foreign tours, they put themselves out of court.

Sentiment or no sentiment, beauty is better than ugliness, and health is better than disease.

Now under the factory system you must sacrifice both health and beauty.

As to my second objection – the evil effect of the factory system on the public health. What are the chief means to health?

Pure air, pure water, pure and sufficient food, cleanliness, exercise, rest, warmth, and ease of mind.

What are the invariable accompaniments of the factory system?

Foul air, foul water, adulterated food, dirt, long hours of sedentary labour, and continual anxiety as to wages and employment in the present, added to a terrible uncertainty as to existence in the future.

Look through any great industrial town in the colliery, the iron, the silk, the cotton, or the woollen industries, and you will find hard work, unhealthy work, vile air, over-crowding, disease, ugliness, drunkenness, and a high death rate. These are facts.

Merrie England, London: Clarion (1894), 19–23.

Lily Gair Wilkinson

Women in freedom

If the terms of slavery are even more ghastly for women than for men, so much the greater must be their effort towards freedom. . . .

A free man or woman is one who can dispose of his or her person without let or hindrance, without reference to any master. If you, being a woman, resolved to be free in this social sense, to go out into the world as a woman in freedom, how would it fare with you?

For a time you might wander unhindered, elated by thoughts of liberty, but very soon you would find that you cannot dwell for ever on the heights. Let us suppose that you feel tired, and that you enter a tea-shop in default of a better place of rest. The shop looks sordid and dingy, and you shudder slightly as a vision of true repose comes to mind – something with green fields and running water and the scent of grass and flowers in it. But, alas! you are not free to that extent; here are no Elysian fields – here is London with its dreary grey buildings and endless discomfort. So you enter the shop. A pale, grim young woman comes up as you choose a seat, and asks what she will bring. You desire only rest, but once more you are reminded that you are not free to choose; rest of a kind you may have, but at the same time tea and buns will be forced upon you. You settle yourself in your uncomfortable corner, sip some of the nasty tea, taste a bun, and ruminate dubiously about your determination to be free. The grim young woman presently brings the bill for tea and cakes, and you realise in a flash that here again in the person of the shop-girl is a limitation of freedom – you are not free from *her*. To the extent that your needs have been satisfied by her service, to this extent your life is dependent upon that service. At this point where you and she have met in life, the one as receiver and the other as the giver of service, each is to a certain degree dependent upon the other.

And in a flash, you recognize the social nature of freedom: how none stands alone in life, but the life of each is dependent upon the lives of others and affected by the lives of others; how the poor are dependent upon the rich, and the rich upon the poor; how the sick are affected by the healthy, and the healthy are affected – or infected – by the sick; how consumers are dependent on producers, and producers on consumers; how the learned are affected by the ignorant, and the ignorant by the learned; and so on throughout the whole range of human relations. And if your vision is clear enough, you realise that so long as one, even the least, of these human brothers and sisters is in bondage, there can be no true freedom for you.

As you pay the bill for tea and cakes, and bid the grim young woman good-day, you have a remembrance perhaps of the feasts in Morris's 'News from Nowhere', when the bearers of food brought along with it, not bills, but roses and kind smiles and friendly words. Alas, again, for freedom!

If your resolve to be free is not quite ended by this illuminating experience in a tea-shop, surely your further experiences must end it soon. Even if circumstances favour you today, tomorrow must put an end to the dream. The sun shines perhaps, the breeze blows, clouds chase each other across the sky. You awake to it all, feeling glad and young and gay and *free*. You resolve to go out into country places where you may be in the companion-ship of free things – flowers and birds and dancing insects. For only one vivid, brilliant day you will be one of the free, you will live as all Nature is calling upon you to live, in idle enjoyment of the sunshine – freedom at least for a day!

But stop! What is that you hear? What is that monotonous beat? It is the clock ticking out the seconds which remain between breakfast and office hours. In half an hour you are due at the office. Now, then, be free for a day if you dare!

Then comes the overwhelming recollection of life as it is; the noise and the crush and the horror of the great city; the strife and labour and feverish competition; disease and death, suffering and starvation. And you see yourself among those who strive and push in the midst of this seething mass of millions of human beings, who hurry hither and thither in frantic efforts to maintain life in enmity with their fellows. You see yourself with nerves strained and brain exhausted, working hour after hour at the hateful machine, to be the human part of which you have sold your living body. For it is not worked by electric power alone, but by human power also.

Dare to be free for a day – and what then?

If you dare to be free for even one day, you will be thrust out by your fellows, another will take your place; the machine will still be served with its due of human energy; this great industrial activity which pollutes the air and obscures the sunlight will not be interrupted for one instant by the want of you – you will not be missed.

But you? The means of life will be gone for you; the price of your freedom will be poverty and death.

In that monster army of modern industrial life the penalty of desertion is death. There is no way of living for you in the wild outside of it. The woods and the fields and the rivers and all the rich, beautiful country all belong to individuals of whom you know nothing and who know nothing of you, who care nothing for you. They will not permit you to take to your use so much earth as may fill a flower-pot – hands off! it is private property! Let the human body perish; the law allows it, and will even provide for you a pauper's grave. But let the sacred rights of private property be in the least degree violated, and the law in all its might is there to do vengeance and give protection to the proprietor.

No, the slave of the industrial system cannot be free for even one day. Turn back quickly to the city again and sell yourself once more into slavery before it is too late.

Here, too, everything belongs to individuals of whom you know nothing and who know nothing of you. All the tremendous machinery by which the few things needful and the many needless are being produced, and the buildings which contain the machinery, and the ground upon which the buildings stand – all belong to these unseen, unknown human beings in possession. And to sell yourself bodily for all the long beautiful hours of your precious days of youth to these possessors is your only means of life.

So once again, as you stand listening to the menace of the clock and wondering whether you will break free or trudge back to the office, you have a sudden revelation. You realise that while there are men and women who hold from others the means of life – the rich surface of the earth and the means of cultivating that richness – so long there will be no freedom for the others who possess none of it at all. For possession by a few gives power to the few to control the lives of the millions who are dispossessed, and to bind them in lifelong bondage.

You have thus arrived at a great illumination through your vain striving after personal liberty.

There can be no freedom for single individuals – one here and one there cannot be free in a social sense; but men and women, being socially interdependent, can only be free together – as a *community*, that is. And further, there can be no freedom while there is private property which prevents all men and women having free access to the means of life; not one here and one there must be possessors, but all must possess together – *in common*, that is.

And this is Communism.

If ever men and women attain these essentials of freedom, the life of human beings will be a Communistic life and the most terrible impediments to a full and true human development may thus be overcome.

How, then, will it fare, especially with women?

Women will have the same freedom as men, because they will be able to dispose of their lives as they choose.

Mrs Wibaut and L. G. Wilkinson, *Women in Rebellion (1900)*, London: Independent Labour Party (1973), 23–8.

CHAPTER 18

Green politics

C. Lytton, Prisons and prisoners
G. Pinchot, The politics of conservation
L. Paul, Angry young man
M. Horkheimer, The revolt of nature

Introduction

> For a while in the 1920s, before they were dislodged by right-wing *coups d'état*, peasant parties with large electoral majorities ruled most of the nations of Eastern Europe. The climax of this false dawn came in 1927 with the founding of the Green International, eventually joined by seventeen European peasant parties, though it collapsed as the 1930s ushered in the age of the dictators.
>
> (Canovan 1977: 88)

Green parties contesting elections, rather than advocating non-violent insurrection, are a very recent phenomenon. The first candidate in Britain to stand on a clear Green platform of decentralization, environmental concern and opposition to economic growth, the Rev. John Papworth, did so in a parliamentary by-election as late as 1970 (Papworth 1971). A surge of environmental concern in the late 1960s and early 1970s, along with the publication of *Limits to Growth* and *Blueprint for Survival*, inspired the creation of 'ecology' parties in Britain, France, New Zealand and Australia (Kemp and Wall 1990). A fully fledged Green manifesto combining ecology with a full statement of social demands became widespread only in the early 1980s, inspired by the initial success of the German Greens.

Earlier political movements have clearly contained Green elements. Most have favoured direct action or the creation of alternative communities rather than parliamentary process. We may talk in this regard, as does E. M. Forster, of the 'socialism of Blake and Shelley', invoking Peter Kropotkin and the Diggers. But we may not talk of a party political Green movement. Whilst Petra Kelly may have hoped that the German Greens would become an 'anti-party party', they have in fact adhered to a largely conventional model of electoral activity and leadership values. Green politics, in that it can be said to have existed prior to the 1970s, has been linked with extra-parliamentary activities. The diversity of these movements is at first sight rather surprising. For example, during the decades between 1920 and 1970 we find an eccentric right wing including Social Credit and the Catholic Distributists, as well as equally eccentric far-left Situationists and Kaboteurs (Gray 1974; Paul 1951; Webb 1981). In the 1930s socialists such as Leslie Paul and Richard Acland shared opinions on the despoliation of the countryside with conservatives such as H. J. Massingham and Rolf Gardiner (Gardiner 1949; Massingham 1941; Penty 1937; Prynn 1972; Stephenson 1989).

Our first extract, from a radical but far from revolutionary Suffragette, Constance Lytton, illustrates concern with Green themes, including local democratic control, pollution, 'the monster of industrialism' and countryside conservation. Our second contributor, the US associate of President Theodore Roosevelt, Gifford Pinchot, is equally a reformer rather than an extra-parliamentiary radical. He argued that conservation (in contrast to radical environmentalism) 'has captured the Nation', describing the contest between democratic forces and special business interests which seek to exploit the Earth and the American people for short-term gain. Pinchot, a former US Chief Forester, was a prominent figure during the Progressive era and was later a presidential candidate (Hays 1959; Worster 1991: 266–8).

Next we find Leslie Paul describing the 'green' orientation of the youth movements of the 1920s. Hargraves, later a founder of the Social Credit Greenshirts, a movement for economic reform based upon the ideas of Major Douglas, attacked Baden-Powell's Scouts as militaristic and anti-democratic. He created his own movement for young adults, the Kibbo Kift Kin, which celebrated an ecological outlook and the American Indian tradition of Seton Thompson (Finlay 1972). In turn Paul, a young supporter of Britain's Labour and co-operative movement, provoked a further split which led to the creation of the Woodcraft Folk, a radical and environmentally conscious scouting body that thrives to this day. With the collapse of the Morris/Carpenter conception of ecological socialism, both the Kift and the Woodcraft Folk were instrumental in maintaining interest in broadly Green ideas (Morris 1970: 193–4). Something of a Green tradition was also maintained with the creation by G. K. Chesterton and Hillaire Belloc of the Distributist League, a movement that during the 1920s and 1930s advocated decentralization, peasant values and the redistribution of land. Its love of medieval institutions, antisemitism and co-option by Mosley's British Union of Fascists pushed the politics of environmental concern, briefly, towards the far right. At the same time an entirely different train of thought was evolving through the work of the Frankfurt school.

After escaping Hitler's Germany for the United States, the leading members of the Frankfurt school of critical theory, Adorno and Horkheimer, fused the politics of ecological concern with Marxism. They feared that the Age of Reason which had allowed huge advances in science and banished superstition had become the instrument of an unreason that made humanity the servant of blind forces. Advances in knowledge had been used merely to enhance the might of a few; progress and technology needed to be reappraised if they were to lead to real benefit. The ambiguous and complex philosophy developed out of such a thesis is perhaps most clearly expressed in Horkheimer's *The Eclipse of Reason*. While right-wing ecologists created the Soil Association in Britain and were active in the US conservation movement, the Frankfurt school directly created the conditions for a fuller and more human Green politics in the 1960s and 1970s. Bookchin and Marcuse, the Situationists, the German Green movement, the American counter-culture and the Dutch Kaboteurs all owe an explicit debt to the Frankfurt school in their attempts to produce a living and ecologically aware politics.

Constance Lytton

Prisons and prisoners

In 1896 and successive years I had given secretarial help to my aunt, Mrs. C. W. Earle, in the writing of her wonderfully delightful books, beginning with 'Pot Pourri from a Surrey Garden'. She insisted that, in return for my small and mostly mechanical services, we should share the profits of the sale. The book ran into many editions, and she held to her bargain, but I never felt as if I had a right to the money. Her widely sympathetic and stimulating companionship had a great influence on my mind. Thanks to her investigations in theories of diet, I became a strict vegetarian. My health gained in all directions and I gradually freed myself from the so-called 'constitutional' rheumatism from which I had suffered since my infancy. I realised, too, that all these years I had caused untold suffering that I might be fed, and determined that in future the unnatural death of an animal should not be necessary to make up my bill of fare. My vitality increased, but the notion of a vocation apart from my family and home remained as foreign to my ideas as it was then to the average British spinster of my class.

In the year 1906 my godmother, Lady Bloomfield, died. She had shown me much kindness and I had never found an opportunity to serve her in any way, the generosity had been all on her side; yet, at her death, she left me some money, without any conditions as to how I should spend it. It gave me a strange new feeling of power and exhilaration. I look back upon this event as being spiritually the starting point in my new life, of which this book will tell, although, from the practical point of view, it seems only by a series of coincidences that my after experiences were evolved from it.

I looked about me with a view to spending the money. I had a fancy to put it to some public use. The commonly accepted channels of philanthropy did not appeal to me. I shifted my inquiries in other directions. I remember that at this time I was chiefly occupied with the idea that reformers were for the most part town dwellers, their philosophy and schemes attuned to those surroundings. There seemed to me need for a counteracting influence to attempt reform and regeneration on behalf of country dwellers. The noiseless revolution which had been worked in a few decades by the system of compulsory education seemed to me tainted throughout by the ideals of townsfolk. The influence of teachers and clergy, of public authorities in general, sets before the nation's children and their arents ideals which mould them into townsfolk. Country craft and country lore grow less, and are less honoured in every decade. There is no room for them in the national curriculum. This tendency, nevertheless, seems unnatural, imposed by a species of force on a reluctant though inarticulate people. Some temperaments cannot acclimatise themselves to town dwelling. The life of cities will always appear to them artificial, repellent from the physical conditions it imposes and the mental outlook resulting from these. Rain, earth and air are better scavengers than any municipal corporation. The ceaseless cleansing, yet never making clean, of town existence has from my childhood fretted my imagination and produced a sense of incarceration. In towns, the earth is laid over with tombstones, metalled roads or floors of wood. Avenues of bricks and mortar shut out a great part of the sky, limiting into a mere ceiling the heavens which should be our surrounding. A town had always seemed to me a 'deterrent' workhouse at best, and often a punitive prison besides.

The monster of industrialism, which followed in the wake of the discovery of steam and the dethroning of handicraft by artificially propelled machinery, may one day be bridled and controlled so as to be a servant of humanity, a fellow worker in the day-to-day glory of creation; but for the present it is still a wild beast, a dragon at large, dealing pestilence and death with its fiery breath, combated in panic, its evils evaded rather than faced, its power a nightmare breeding fear and subjection. Instead of harnessing this new force to every branch of our existence, ordering it to serve us at our command, we have cringed before it, left our normal lives and drained our energies to congregate in its grimy temples and workshop at its shrine. Poor, blind force that it is, we are determined to make it an idol, and for the sake of the return in money which its mechanical rotations produce, we have been willing to sacrifice the interests both of the human beings which should control it and of the soil, the land, which alone can produce the raw material for its task.

How to transform this Moloch from a tyrannous master to a helpful, submissive friend, that was the problem which seemed to cry out for solution above all others. I looked around to see how the needs of country folk were able to express themselves, and everywhere there was presented to my inquiries a complicated machinery of administration both national and local – voting rights for election of parliamentary, municipal, county, district and parish councils. But this machinery was apparently born and bred of urban conditions, superimposed upon the rural districts, in no way native to them, not of spontaneous growth. This succession of councils, instituted with the apparent purpose of watching one another and, if necessary, bringing pressure to bear upon one another, were for the most part lifeless formalities, having no organic life, no breath of reality to set them in motion. The only function which gives them any tangible vitality is their power of imposing taxation and levying rates. . . .

Prisons and Prisoners, London: Heinemann (1914), 2–5.

Gifford Pinchot

The politics of conservation

Conservation has captured the Nation. Its progress during the last twelve months is amazing. Official opposition to the Conservation Movement, whatever damage it has done or still threatens to the public interest, has vastly strengthened the grasp of Conservation upon the minds and consciences of our people. Efforts to observe or belittle the issue have only served to make it larger and clearer in the public estimation. The Conservation movement cannot be checked by the baseless charge that it will prevent development, or that every man who tells the plain truth is either a muck-raker or a demagogue. It has taken firm hold on our national, moral sense, and when an issue does that it has won. . . .

Because the special interests are in politics, we as a Nation have lost confidence in Congress....

For a dozen years the demand of the Nation for the Pure Food and Drug Bill was outweighed in Congress by the interests which asserted their right to poison the people for a profit.

Congress refused to authorize the preparation of a great plan of waterway development in the general interest, and for ten years has declined to pass the Appalachian and White Mountain National Forest Bill, although the people are practically unanimous for both....

A representative of the people who wears the collar of the special interests has touched bottom. He can sink no further....

This Nation has decided to do away with government by money for profit and return to the government our forefathers died for and gave to us – government by men for human welfare and human progress....

The Conservation of political liberty will take its proper place alongside the Conservation of the means of living, and in both we shall look to the permanent welfare of the plain people as the supreme end. The way out lies in direct interest by the people in their own affairs and direct action in the few great things that really count.

What is the conclusion of the whole matter? The special interests must be put out of politics. I believe the young men will do it.

The Fight for Conservation, New York: Harcourt Brace (1901), 133, 134, 136, 137, 146, 147.

Leslie Paul

Angry young man

In the Scout movement, with its millions of impressionable boys and young men, organized on a world-wide basis, Hargrave's future had seemed utterly assured. Yet he chose to break with it, on the two issues that troubled the conscience of all of us, war and democracy. There was, in the Scout movement in those days, no means by which the rank-and-file could influence the policy or change the leadership of the movement. And on the issue of *militarism*, the pre-war record of the Scout movement was not a very happy one. It was because Hargrave demanded reforms in these matters, including a more democratic constitution, that he was expelled. And when he launched his new movement, Kibbo Kift, or the 'Proof of Great Strength', its declaration of aims, called 'The Covenant of the Kibbo Kift', was like a new wind blowing through our country. Under the influence of H. G. Wells, it spoke strongly for peace, world unity and world government: from Wells too came the conception of a New Samurai which inspired it. The Covenant asked for co-operative woodland communities and the revival of native arts and crafts, the restoration of rural industries and the renewal, through a new education on woodcraft lines, of the old folk life of the people, now buried under a machine-made civilization. Like Edward Carpenter, we looked upon civilization as a disease our new movement was to cure: if we visualized a new society it must have approximated to that which William Morris describes in *News from Nowhere*. The advocacy of Craft Guilds gave encouragement to the socialist elements in the new movement. The demand for land reservations and national parks breathed the hope of salvation from endless urbanization. And there was, above all, the direct and unforgettable command to members to seek pride of body, balance of mind and vital spiritual perception. In those days of hope we were carried away.

Hargrave brought his sense of form to bear upon the costume and practices of the Kibbo Kift. We dressed in cowled jerkins and wore shorts. Our leather belts were handmade, our badges hand-decorated and Hargrave's own excellent designs saved us from the inept and the banal. We loaded our gear into handsome rucsacs and tramped with rough ash staves in hand. The simple, archaic monkish costume seemed itself to witness to the rougher, more self-reliant, yet more brotherly life we were going to pursue with a religious devotion. The development of a campfire ritual of extraordinary splendour – the celebrants of this strange mass wore embroidered robes, and intoned a liturgy to the swinging of censers as they lighted the ceremonial fire – promised the birth of a new, pagan religion. We were certain that we were the new elite, and that by some mystical process we had been chosen to transform the world.

My exhilarating promotion to be chief of a paper devoted to 'Rovering, camping and woodcraft' gave me a professional interest in youth movements, as well as acquaintance with many youth leaders. The life of youth movements I conceived to be my true business, to which even my writing had to take second place, and all my spare time was spent at the camps and gatherings of the lodges. I had founded my own youth lodge, of which Roly, and Eric Greenhill, the actor, were the principal members. Gorden Ellis, who then lived in Deptford, put up the idea that we should bring together all the scattered members of John Hargrave's movement in South East London and form them into a local association. The first meeting was held in the mayor's parlour at Deptford Town Hall and among those present were John Wilmot, who had unsuccessfully contested East Lewisham as a Labour candidate and so fired the opening shots of his political career, Joseph Reeves, who was running Co-operative education locally, a man of intense energy and imagination, and many others prominent in South London socialism. To my surprise I was appointed leader, an honour I tried to refuse and hardly knew how to sustain. By a series of mischances this quite premature promotion of mine was to be the cause of a serious split in Kibbo Kift, which had the consequence of destroying it.

This election took place about the time that I took over *The Open Road*: it was about then, too, that I went with friends to visit John Hargrave in his little bungalow at Kings Langley. We pitched our one-man tents in the waste land behind his bungalow and came in, when summoned, to sit at his feet and listen to his omniscient talk. . . .

I came to the conclusion that no effort would be made to establish the new youth movement unless I made it myself. However, I was only nineteen, and I first had to demonstrate to myself that I was capable of doing what I wanted. There was a thoughtful and modest friend of mine, Sidney Shaw, who had been trained as an engineer but was then out of work. With him I planned a small experimental group of boys and girls in which, I said, we would test out afresh what all of us then called 'tribal training' theories of education. Ad if it did well, a new movement, released from the stale debate which had ruined Kibbo Kift, might spring from our efforts. We started early in 1925 in Lewisham, with four small boys: the small boys have long since grown up and married, but they live in the neighbourhood still, and I still hear of them: presently we added small girls so that the movement could be genuinely co-educational. It had long been one of our points of criticism of the Boy Scouts and Girl Guides that the sexes were too sharply separated: one could not have a movement, close in feeling to the family or tribe, without freedom for both sexes within it.

The exhilarating business of building everything, from the ground up, gripped me from the moment of that beginning, and before long a small but genuine movement came into existence which called itself *The Woodcraft Folk* – it used the word in the German sense of *volk* and not in the English 'fairy' or 'art-and-crafty' sense – which was able to rely upon co-operative societies for support and encouragement. Eventually my group and several others came together and drew up a dignified Charter which read:

'We declare that it is our desire to develop in ourselves, for the service of the people, mental and physical health, and communal responsibility, by camping out and living in close contact with nature, by using the creative faculty both of our minds and our hands, and by sincerity in all our dealings with our neighbours: we declare that it is our desire to make ourselves familiar with the history of the world, and the development of man in the slow march of evolution that we may understand and revere the Great Spirit which urges all things to perfect themselves.

'We further declare that the welfare of the community can be assured only when the instruments of production are owned by the community, and all things necessary for the good of the race are produced by common service for the common use; when the production of all things that directly or indirectly destroy human life ceases to be; and when man shall turn his labour from private greed to social service to increase the happiness of mankind, and when nations shall cease to suckle tribal enmities and unite in common fellowship.'

Not long after this, in a pamphlet which drafted an educational programme for the new movement, I wrote this:

'Our education is not a matter of little moral talks or stilted lectures, it is a system wherein the primal instincts of the child are moulded along a social path by the very things a child loves. In truth we let a child train itself and we see that it grows from within and is not coerced from without. Hand in hand we go with our children to explore and examine all that life holds out to us. So it comes about that our principles of training permeate the whole of our activities; every symbol, every totem, every song has its own peculiar value, and no action from acting a charade to boiling a billy of water is devoid of significance. We feel that it is necessary, if the race is to survive, to produce men and women who by their knowledge, their physical fitness and their mental independence shall bring quick, sure brains and boundless vitality to bear on man's struggle for liberty. We are the revolution. With the health that is ours and with the intellect and physique that will be the heritage of those we train we are paving the way for that reorganization of the economic system which will mark the rebirth of the human race.'

I see now a clear contradiction, not obvious to me at the time, between the simple and indeed humble educational argument of the first part of that statement, and the extravagant assertion that we were the revolution. How genuine, however, was our belief that one had to 'Learn by doing, teach by being'. Rousseau's *Emile* had enormous influence on us. . . .

Angry Young Man, London: Faber (1951), 55–6, 62–3.

Martin Horkheimer

The revolt of nature

Intellectually, modern man is less hypocritical than his forefathers of the nineteenth century who glossed over the materialistic practices of society by pious phrases about idealism. Today no one is taken in by this kind of hypocrisy. But this is not because the contradiction between high-sounding phrases and reality has been abolished. The contradiction has only become institutionalized. Hypocrisy has turned cynical; it does not even expect to be believed. The same voice that preaches about the higher things of life, such as art, friendship, or religion, exhorts the hearer to select a given brand of soap. Pamphlets on how to improve one's speech, how to understand music, how to be saved, are written in the same style as those extolling the advantages of laxatives. Indeed, one expert copywriter may have written any one of them. In the highly developed division of labor, expression has become an instrument used by technicians in the service of industry. A would-be author can go to a school and learn the many combinations that can be contrived from a list of set plots. These schemes have been co-ordinated to a certain degree with the requirements of other agencies of mass culture, particularly those of the film industry. A novel is written with its film possibilities in mind, a symphony or poem is composed with an eye to its propaganda value. Once it was the endeavor of art, literature, and philosophy to express the meaning of things and of life, to be the voice of all that is dumb, to endow nature with an organ for making known her sufferings, or, we might say, to call reality by its rightful name. Today nature's tongue is taken away. Once it was thought that each utterance, word, cry, or gesture had an intrinsic meaning; today it is merely an occurrence.

The story of the boy who looked up at the sky and asked, 'Daddy, what is the moon supposed to advertise?' is an allegory of what has happened to the relation between man and nature in the era of formalized reason. On the one hand, nature has been stripped of all intrinsic value or meaning. On the other, man has been stripped of all aims except self-preservation. He tries to transform everything within reach into a means to that end. Every word or sentence that hints of relations other than pragmatic is suspect. When a man is asked to admire a thing, to respect a feeling or attitude, to love a person for his own sake, he smells sentimentality and suspects that someone is pulling his leg or trying to sell him something. . . .

Modern insensitivity to nature is indeed only a variation of the pragmatic attitude that is typical of Western civilization as a whole. The forms are different. The early trapper saw in the prairies and mountains only the prospects of good hunting; the modern business-man sees in the landscape an opportunity for the display of cigarette posters. The fate of animals in our world is symbolized by an item printed in newspapers of a few years ago. It reported that landings of planes in Africa were often hampered by herds of elephants and other beasts. Animals are here considered simply as obstructors of traffic. This mentality of man as the master can be traced back to the first chapters of Genesis. The few precepts in favor of animals that we encounter in the Bible have been interpreted by the most outstanding religious thinkers, Paul, Thomas Aquinas, and Luther, as pertaining only to the moral education of man, and in no wise to any obligation of man toward other creatures. Only man's soul can be saved; animals have but the right to suffer. 'Some men and women,' wrote a British churchman a few years ago, 'suffer and die for the life, the welfare, the happiness of others. This law is continually seen in operation. The supreme example of it was shown to the world (I write with reverence) on Calvary. Why should animals be exempted from the operation of this law or principle?'[6] Pope Pius IX did not permit a society for the prevention of cruelty to animals to be founded in Rome because, as he declared, theology teaches that man owes no duty to any animal.[7] National Socialism, it is true, boasted of its protection of animals, but only in order to humiliate more deeply those 'inferior races' whom they treated as mere nature.

These instances are quoted only in order to show that pragmatic reason is not new. Yet, the philosophy behind it, the idea that reason, the highest intellectual faculty of man, is solely concerned with instruments, nay, is a mere instrument itself, is formulated more clearly and accepted more generally today than ever before. The principle of domination has become the idol to which everything is sacrificed.

The history of man's efforts to subjugate nature is also the history of man's subjugation by man. The development of the concept of the ego reflects this twofold history.

It is very hard to describe precisely what the languages of the Western world have at any given time purported to connote in the term ego – a notion steeped in vague associations. As the principle of the self endeavoring to win in the fight against nature in general, against other people in particular, and against its own impulses, the ego is felt to be related to the functions of domination, command, and organization. The ego principle seems to be manifested in the outstretched arm of the ruler, directing his men to march or dooming the culprit to execution. Spiritually, it has the quality of a ray of light. In penetrating the darkness, it startles the ghosts of belief and feeling, which prefer to lurk in the shadows. Historically, it belongs pre-eminently to an age of caste privilege marked by a cleavage between intellectual and manual labour, between conquerors and conquered. Its dominance is patent in the patriarchal epoch. . . .

The entire universe becomes a tool of the ego, although the ego has no substance or meaning except in its own boundless activity. Modern ideology, though much closer to Fichte than is generally believed, has cut adrift from such metaphysical moorings, and the antagonism between an abstract ego as undisputed master and a nature stripped of inherent meaning is obscured by vague absolutes such as the ideas of progress, success, happiness, or experience.

Nevertheless, nature is today more than ever conceived as a mere tool of man. It is the object of total exploitation that has no aim set by reason, and therefore no limit. Man's boundless imperialism is never satisfied. The dominion of the human race over the earth has no parallel in those epochs of natural history in which other animal species represented the highest forms of organic development. Their appetites were limited by the necessities of their physical existence. Indeed, man's avidity to extend his power in two infinities, the microcosm and the universe, does not arise directly from his own nature, but from the structure of society. Just as attacks of imperialistic nations on the rest of the world must be explained on the basis of their internal struggles rather than in terms of their so-called national character, to so the totalitarian attack of the human race on anything that it excludes from itself derives from interhuman relationships rather than from innate human qualities. The warfare among men in war and in peace is the key to the insatiability of the species and to its ensuing practical attitudes, as well as to the categories and methods of scientific intelligence in which nature appears increasingly under the aspect of its most effective exploitation. This form of perception has also determined the way in which human beings visualize each other in their economic and political relationships.

Notes

6   Edward Westermark, *Christianity and Morals*, New York, 1939, p. 388.
7   ibid., p. 389.

The Eclipse of Reason, New York: Oxford University Press (1947), 100–1, 104–5, 108–9.

CHAPTER 19

Utopia or else!

W. H. Hudson, *A Crystal Age*

W. Morris, *News from Nowhere*

C. P. Gilman, *Our growing modesty*

A. Huxley, *The island of Pala*

R. Dumont, *Manifesto of an alternative culture*

Introduction

> A map of the world that does not include Utopia is not even worth glancing at, for it leaves out the one country at which Humanity is always landing ... Progress is the realization of Utopia.
>
> (Oscar Wilde 1987: 3)

René Dumont, the Ecology candidate in the 1974 French presidential election, entitled his book on the environmental crisis *Utopia or Else*. The concept of 'utopia', variously translated as 'nowhere' (*outopia*) or 'perfection' (*eutopia*), has powerfully inspired the Green movement. Greens would argue that to solve ecological problems requires the transformation both of institutions and of the individual, resulting in the creation of a new society. The construction of this vision of a better world inspires us to change the society we have and clarifies criticism of what exists in the present. Refinement of the 'utopian' model allows for the construction of better alternatives and also acts as a motivating myth. Ecological harmony, social justice, spiritual values and sexual liberation are found in the utopian tradition stretching from Plato's description of Atlantis and More's *Utopia* to Callanbach's *Ecotopia*, written in the 1970s. Dystopias, or anti-utopias, have also been produced, Aldous Huxley's *Brave New World* supplying powerful criticism of existing expansionist trends.

The naturalist W. H. Hudson (1841–1922) created an ecotopia in his sentimental tale *A Crystal Age*. Although unexceptionally crafted, it deals with a number of environmental themes. Here we find the hero, who has been transmitted into an ecologically sustainable future society, sampling his first vegetarian meal and debating the ethics of mushrooms and milk. He is told of how the technological exploitation of his own Victorian century gave rise to hubris and disaster. 'Their vain ambition lasted on, and the end of it was death.' Morris (1834–96), in *News from Nowhere*, provides a more detailed and literary portrait of the future. Here, we are told, is a post-political society without nations or government. The reference to 'Horrebow's Snakes in Iceland' is clear; there are no snakes in Iceland and the old conflicts are equally absent in Morris's world, where money has been abolished, production functions on a small, localized scale and salmon swim up the Thames.

Charlotte Perkins Gilman, as in Chapter Six, provides us with the example of a feminist ecotopia in *Herland*. It is the earliest example of a genre continued by women science fiction writers such as Ursula Le Guin and Marge Piercy. Anticipating the modern method of permaculture, the women of Herland replant entire forests as a means of feeding themselves with minimum effort. Their organic and balanced economy is described in detail; 'All the scraps and leavings of their food, plant waste from lumber work or textile industry, all the solid matter from the sewage, properly treated and combined – everything which came from the earth went back into it.'

Next we reach Huxley's comprehensive, but in a didactic genre, overly preaching, description of *Island*. In this, his last novel, he constructs an alternative to the hellish super-industrialization of *Brave New World*. Ecological management, economics, education, energy production, health, religion and many other matters receive the Huxley treatment. Lord Edward, whom we left attacking progress in Chapter Nine, would have loved it. The fruits of such utopianism are well illustrated in our last passage. *Manifesto of an Alternative Culture*, published by René Dumont in support of his candidacy in the French elections, provides one of the earliest modern examples of a fully fledged Green programme, linking the themes outlined in this volume.

W. H. Hudson

A Crystal Age

Now I am not at all particular about what I eat, as with me a good digestion waits on appetite, and as long as I get a bellyful – to use a good old sensible English word – I am satisfied. On this particular occasion I could have consumed a haggis without assistance from anyone. I felt so ravenously hungry: yet I take the haggis to be the greatest culinary abomination ever invented by flesh-eating barbarians. I was therefore not a little disappointed when nothing more substantial than a dish of some whity-green, crisp-looking stuff, resembling endive, was placed before me by one of the picturesque handmaidens. Of course politeness, to say nothing of my raving appetite, I found it both cold and bitter-tasting, although the dish on which it was served was extremely beautiful to the eye. When I had devoured the last green leaf I felt hungrier than ever and began to wonder whether it would be right to ask for more when, to my unspeakable relief and gratification, other more succulent dishes followed. Bruised grain and pulse were the principal ingredients in some of them, and they were certainly very nice. We also had some pleasant tasting beverages, made, I believe, from the juice of various fruits, but the delicious alcoholic sting was in none of them. No animal food appeared except milk, which was in some of the stews or soups. But then milk is perhaps not an animal food at all. At all events, I have heard vegetarians say that it is nothing more than the juices of certain plants which have undergone some chemical change in the cow's system. On the other hand, I have known others fly to the opposite extreme and denounce mushrooms as belonging to a class of things low down at the root of all life, where animal and vegetable are not easily distinguishable. But these are matters I know nothing about. The repast concluded with fruits, well flavoured but unfamiliar to me, though there was one surprisingly like a peach, only the stone was no bigger than a cherry. There was also a very delicious little confection of crushed nuts and honey. . . .

241

Thus we know that in the past men sought after knowledge of various kinds, asking not whether it was for good or for evil; but every offence of the mind and the body has its appropriate reward; and while their knowledge grew apace, that better knowledge and discrimination which the Father gives in every living soul, both in man and in beast, was taken from them. Thus by increasing their riches they were made poorer; and like one who, forgetting the limits that are set by the faculties, gazes steadfastly on the sun, staring so much they become afflicted with blindness. Thus did they thirst, and drink again, and were crazed; being inflamed with the desire to learn secrets of nature, and hesitating not to dip their hands in blood, seeking in the living tissues of animals for the hidden springs of life. For in their madness they hoped by knowledge to gain absolute domination over nature, thereby taking from the Father of the world his prerogative.

But their vain ambition lasted not, and the end of it was death.

A Crystal Age, London: Fisher Unwin (1877), 45–6, 69–70.

William Morris

News from Nowhere

Chapter XIII. Concerning politics

Said I: 'How do you manage with politics?'

Said Hammond, smiling: 'I am glad that it is of *me* that you ask that question; I do believe that anybody else would make you explain yourself, or try to do so, till you were sickened of asking questions. Indeed, I believe I am the only man in England who would know what you mean; and since I know, I will answer your question briefly by saying that we are very well off as to politics, – because we have none. If ever you make a book out of this conversation, put this in a chapter by itself, after the model of old Horrebow's Snakes in Iceland.'

'I will,' said I.

Chapter XIV. How matters are managed

Said I: 'How about your relations with foreign nations?"

'I will not affect not to know what you mean,' said he, 'but I will tell you at once that the whole system of rival and contending nations which played so great a part in the "government" of the world of civilisation has disappeared along with the inequality betwixt man and man in society.'

'Does not that make the world duller?' said I.

'Why?' said the old man.

'The obliteration of national variety,' said I.

'Nonsense,' he said, somewhat snappishly. 'Cross the water and see. You will find plenty of variety: the landscape, the building, the diet, the amusements, all various. The men and women varying in looks as well as in habits of thought; the costume far more various than in the commercial period. How should it add to the variety or dispel the dulness, to coerce certain families or tribes, often heterogeneous and jarring with one another, into certain artificial and mechanical groups, and call them nations, and stimulate their patriotism – *i.e.*, their foolish and envious prejudices?'

'Well – I don't know how,' said I.

'That's right,' said Hammond cheerily; 'you can easily understand that now we are freed from this folly it is obvious to us that by means of this very diversity the different strains of blood in the world can be serviceable and pleasant to each other, without in the least wanting to rob each other: we are all bent on the same enterprise, making the most of our lives. And I must tell you, whatever quarrels or misunderstandings arise, they very seldom take place between people of different race; and consequently since there is less unreason in them, they are the more readily appeased.'

'Good,' said I, 'but as to those matters of politics; as to general differences of opinion in one and the same community. Do you assert that there are none?'

'No, not at all,' said he, somewhat snappishly; 'but I do say that differences of opinion about real solid things need not, and with us do not, crystallise people into parties permanently hostile to one another, with different theories as to the build of the universe and the progress of time. Isn't that what politics used to mean?'

'H'm, well,' said I, 'I am not so sure of that.'

News from Nowhere (1890), London, Longman (1907), 94–5.

Charlotte Perkins Gilman

Our growing modesty

Having improved their agriculture to the highest point, and carefully estimated the number of persons who could comfortably live on their square miles; having then limited their population to that number, one would think that was all there was to be done. But they had not thought so. To them the country was a unit – it was theirs. They themselves were a unit, a conscious group; they thought in terms of the community. As such, their time-sense was not limited to the hopes and ambitions of an individual life. Therefore, they habitually considered and carried out plans for improvement which might cover centuries.

I had never seen, had scarcely imagined, human beings undertaking such a work as the deliberate replanting of an entire forest area with different kinds of trees. Yet this seemed to them the simplest common sense, like a man's plowing up an inferior lawn and reseeding it. Now every tree bore fruit – edible fruit, that is. In the case of one tree, in which they took especial pride, it had originally no fruit at all – that is, none humanly edible – yet was so beautiful that they wished to keep it. For nine hundred years they had experimented, and now showed us this particularly lovely graceful tree, with a profuse crop of nutritious seeds.

They had early decided that trees were the best food plants, requiring far less labor in tilling the soil, and bearing a larger amount of food for the same ground space; also doing much to preserve and enrich the soil.

Due regard had been paid to seasonable crops, and their fruit and nuts, grains and berries, kept on almost the year through.

On the higher part of the country, near the backing wall of mountains, they had a real winter with snow. Toward the southeastern point, where there was a large valley with a lake whose outlet was subterranean, the climate was like that of California, and citrus fruits, figs, and olives grew abundantly.

What impressed me particularly was their scheme of fertilization. Here was this little shut-in piece of land where one would have thought an ordinary people would have been starved out long ago or reduced to an annual struggle for life. These careful culturists had worked out a perfect scheme of refeeding the soil with all that came out of it. All the scraps and leavings of their food, plant waste from lumber work or textile industry, all the solid matter from the sewage, properly treated and combined – everything which came from the earth went back to it.

The practical result was like that in any healthy forest; an increasingly valuable soil was being built, instead of the progressive impoverishment so often seen in the rest of the world.

When this first burst upon us we made such approving comments that they were surprised that such obvious common sense should be praised; asked what are methods were; and we had some difficulty in – well, in diverting them, by referring to the extent of our own land, and the – admitted – carelessness with which we had just skimmed the cream of it.

At least we thought we had diverted them. Later I found that besides keeping a careful and accurate account of all we told them, they had a sort of skeleton chart, on which the things we said and the things we palpably avoided saying were all set down and studied. It really was child's play for those profound educators to work out a painfully accurate estimate of our conditions – in some lines. When a given line of observation seemed to lead to some very dreadful inference they always gave us the benefit of the doubt, leaving it open to further knowledge. Some of the things we had grown to accept as perfectly natural, or as belonging to our human limitations, they literally could not have believed; and, as I have said, we had all of us joined in a tacit endeavor to conceal much of the social status at home.

'Confound their grandmotherly minds!' Terry said. 'Of course they can't understand a Man's World! They aren't human – they're just a pack of Fe-Fe-Females!' This was after he had to admit their parthenogenesis.

'I wish our grandfatherly minds had managed as well,' said Jeff. 'Do you really think it's to our credit that we have muddled along with all our poverty and disease and the like? They have peace and plenty, wealth and beauty, goodness and intellect. Pretty good people, I think!'

'You'll find they have their faults too,' Terry insisted; and partly in self-defense, we all three began to look for those faults of theirs. We had been very strong on this subject before we got there – in those baseless speculations of ours.

'Suppose there is a country of women only,' Jeff had put it, over and over. 'What'll they be like?'

And we had been cocksure as to the inevitable limitations, the faults and vices, of a lot of women. We had expected them to be given over to what we called 'feminine vanity' – 'frills and furbelows', and we found they had evolved a costume more perfect than the Chinese dress, richly beautiful when so desired, always useful, of unfailing dignity and good taste.

We had expected a dull submissive monotony, and found a daring social inventiveness far beyond our own, and a mechanical and scientific development fully equal to ours.

We had expected pettiness, and found a social consciousness besides which our nations looked like quarrelling children – feebleminded ones at that.

We had expected jealousy, and found a broad sisterly affection, a fair-minded intelligence, to which we could produce no parallel.

We had expected hysteria, and found a standard of health and vigor, a calmness of temper, to which the habit of profanity, for instance, was impossible to explain – we tried it.

All these things even Terry had to admit, but he still insisted that we should find out the other side pretty soon.

'It stands to reason, doesn't it?' he argued. 'The whole thing's deuced unnatural! – I'd say impossible if we weren't in it. And an unnatural condition's sure to have unnatural results. You'll find some awful characteristics – see if you don't! For instance – we don't know yet what they do with their criminals – their defectives – their aged. You notice we haven't seen any! There's got to be something!'

I was inclined to believe that there had to be something, so I took the bull by the horns – the cow, I should say! – and asked Somel.

'I want to find some flaw in all this perfection,' I told her flatly. 'It simply isn't possible that three million people have no faults. We are trying our best to understand and learn – would you mind helping us by saying what, to your minds, are the worst qualities of this unique civilisation of yours?'

We were sitting together in a shaded arbor, in one of those eating-gardens of theirs. The delicious food had been eaten, a plate of fruit still before us. We could look out on one side over a stretch of open country, quietly rich and lovely . . .

Herland (1915), London: Women's Press (1979), 79–81

Aldous Huxley

The island of Pala

'I'm sorry we can't provide more comfortable transportation,' said Vijaya as they bumped and rattled along.

Will patted Murugan's knee. 'This is the man you should be apologizing to,' he said. 'The one whose soul yearns for Jaguars and Thunderbirds.'

'It's a yearning, I'm afraid,' said Dr Robert from the back seat, 'that will have to remain unsatisfied.'

Murugan made no comment, but smiled the secret contemptuous smile of one who knows better.

'We can't import toys,' Dr Robert went on. 'Only essentials.'

'Such as?'

'You'll see in a moment.' They rounded a curve, and there beneath them were the thatched roofs and tree-shaded gardens of a considerable village. Vijaya pulled up at the side of the road and turned off the motor. 'You're looking at New Rothamsted,' he said. 'Alias Madalia. Rice, vegetables, poultry, fruit. Not to mention two potteries and a furniture factory. Hence those wires.' He waved his hand in the direction of the long row of pylons that climbed up the terraced slope behind the village, dipped out of sight over the ridge, and reappeared, far away, marching up from the floor of the next valley towards the green belt of mountain jungle and the cloudy peaks beyond and above. 'That's one of the indispensable imports – electric equipment. And when the waterfalls have been harnessed and you've strung up the transmission lines, here's something else with a high priority.' He directed a pointing finger at a windowless block of cement that rose incongruously from among the wooden houses near the upper entrance to the village.

'What is it?' Will asked. 'Some kind of electric oven?'

'No, the kilns are over on the other side of the village. This is the communal freezer.'

'In the old days,' Dr Robert explained, 'we used to lose about half of all the perishables we produced. Now we lose practically nothing. Whatever we grow is for us, not for the circumambient bacteria.'

'So now you have enough to eat.'

'More than enough. We eat better than any other country in Asia, and there's a surplus for export. Lenin used to say that electricity plus socialism equals communism. Our equations are rather different. Electricity minus heavy industry plus birth control equals democracy and plenty. Electricity plus heavy industry minus birth control equals misery, totalitarianism and war.'

'Incidentally,' Will asked, 'who owns all this? Are you capitalists or state socialists?'

'Neither. Most of the time we're co-operators. Palanese agriculture has always been an affair of terracing and irrigation. But terracing and irrigation call for pooled efforts and friendly agreements. Cut-throat competition isn't compatible with rice-growing in a mountainous country. Our people found it quite easy to pass from mutual aid in a village community to streamlined co-operative techniques for buying and selling and profit-sharing and financing.'

'Even co-operative financing?'

Dr Robert nodded. 'None of those blood-sucking usurers that you find all over the Indian countryside. And no commercial banks in your Western style. Our borrowing and lending system was modelled on those credit unions that Wilhelm Raiffeisen set up more than a century ago in Germany. Dr Andrew persuaded the Raja to invite one of Raiffeisen's young men to come here and organize a co-operative banking system. It's still going strong.'

'And what do you use for money?' Will asked.

Dr Robert dipped into his trouser pocket and pulled out a handful of silver, gold and copper.

'In a modest way,' he explained, 'Pala's a gold-producing country. We mine enough to give our paper a solid metallic backing. And the gold supplements our exports. We can pay spot cash for expensive equipment like those transmission lines and the generators at the other end.'

'You seem to have solved your economic problems pretty successfully.'

'Solving them wasn't difficult. To begin with, we never allowed ourselves to produce more children than we could feed, clothe, house and educate into something like full humanity. Not being over-populated, we have plenty. But although we have plenty, we've managed to resist the temptation that the West has now succumbed to – the temptation to over-consume. We don't give ourselves coronaries by guzzling six times as much saturated fat as we need. We don't hypnotize ourselves into believing that two television sets will make us twice as happy as one television set. And finally we don't spend a quarter of the gross national product preparing for World War III or even World War's baby brother, Local War MMMCCXXXIII. Armaments, universal debt and planned obsolescence – those are the three pillars of Western prosperity. If war, waste, and moneylenders were abolished, you'd collapse. And while you people are over-consuming, the rest of the world sinks more and more deeply into chronic disaster. Ignorance, militarism and breeding, these three – and the greatest of these is breeding. No hope, not the slightest possibility, of solving the economic problem until *that*'s under control. As population rushes up, prosperity goes down.' He traced the descending curve with an outstretched finger. 'And as prosperity goes down, discontent and rebellion' (the forefinger moved up again), 'political ruthlessness and one-party rule, nationalism and bellicosity begin to rise. Another ten or fifteen years of uninhibited breeding, and the whole world, from China to Peru via Africa and the Middle East will be fairly crawling with Great Leaders, all dedicated to the suppression of freedom, all armed to the teeth by Russia or America or, better still, by both at once, all waving flags, all screaming for *lebensraum*.'

Island, London: Chatto & Windus (1962), 144–6.

René Dumont

Manifesto of an alternative culture

The ecological problem

It is one and the same system which organises the exploitation of the workers and the degradation of living and working conditions and puts the whole earth in danger. The blind policy of growth, which is so extravagantly praised by all the political parties, takes no account either of human well-being or of the environment. In this system the costs of pollution, then of depollution, are added together to swell the production figures, though in fact they cancel one another out. Goods with built-in obsolescence that deteriorate as soon as they are bought, the wastes that accumulate, the production of armaments, the recourse to ever larger and more dangerous technology: our system has to run faster and faster in order to stay where it is.

Nuclear power stations require so much energy for their construction that it is necessary to keep building new ones in preparation for future stations.

In support of this project governments invoke the mystique of progress. Let us be clear on this point: progress whose price is so heavy, for our health, for our children, for the workers, is not progress. Growth has not done away with inequalities in France: it has accentuated them.

On the contrary, a privileged minority benefit from this growth and carefully preserve for themselves an agreeable way of life. All decisions are concentrated in their hands. Centralised control is extended into all spheres and transforms the people who are deprived of information into robots for production and consumption. In this system women have no rights and no voice even in the disposal of their own bodies in the matter of contraception and abortion. In this system a Breton has not the right to be a Breton. Regional cultures are suppressed, uniformity is the rule.

Our 'expansion' has been brought about largely through the pillage of the third world, through under-payment for raw materials, including oil until 1971. This pillage has made possible our unparalleled wastage of all these resources. The famine is due to the breaking down of traditional customs, grain reserves, and irresponsible export of cultures. It is also caused by the extravagant spending of the elites who want to live in western style at the expense of the agricultural and industrial equipment of their countries.

There are solutions

- The primacy of well-being over the accumulation of goods, and the quality of life over the standard of living.
- equilibrium between production, consumption, population and resources.
- transference to the whole population, men and women, within the framework of their communities, of the power to organise themselves, make their own decisions, as well as the power to acquire the necessary information.
- respect for technical and cultural diversity, of human beings and of social groups.
- the use of decentralised production techniques, non-polluting and based on renewable resources, such as, for example, solar energy (soft technology).
- decentralisation of power at all geographical levels (regions, departments, communes, *quartiers*).
- obligatory information to associations about the decisions which concern them, and access to the decision-making procedures.
- the possibility of legal intervention by the associations before the harmful projects are begun.
- the setting up of local means of communication which will allow everyone to express their views and effectively make decisions (local television).

All the present economic calculations are false. They count as an addition to the national wealth expenditure on medicines, costs of hospitalisation, charges for car repairs and costs of burial.

Equally monumental errors today remove all significance from the Gross National

Product (GNP) which is still the official index of progress. A more accurate measure of the national well-being is urgently needed. This would make it possible to escape from the short-term economic perspective based on appetite and power.

To avoid economic crisis and unemployment

- By social measures such as reduction of working hours and rates of working.
- By social investments (crèches, hospitals), the most productive of all.
- By changing industrial production to more durable, useful and less polluting products. This is especially possible for the motor car.
- By altering agricultural policy so as not to favour the moneylenders. Confining of subsidies to activities which do not destroy the natural equilibria.
- By reorientation and development of services such as preventive medicine, state education, permanent citizenship to foreign workers, the protection of nature and the struggle against pollution.

The redistribution of wealth

Democratising education, increasing low wages, helping the aged … is not enough. The redistribution of wealth involves above all a move towards:

- greater equality in the conditions and environment of work, housing and health.
- greater equality for all in quality and standard of life – a fairer relationship between the prices of agricultural and industrial products.
- financially it requires a complete rethinking of the distribution of the national wealth.
- giving the major part of pubic money to the local communities.
- economising by avoiding waste.
- giving priority to social spending for the betterment of the environment of the under-privileged.
- a generalised tax on pollution.

But don't wait for things to change by themselves. Only you have the power to change them.

'Manifesto of an alternative culture', *Undercurrents* 10 (1975), 6.

Suggested further reading

The titles listed here may be used to place extracts in context and to cast some light on areas of controversy. I have tried to emphasize areas where little has been written and exploration may seem more particularly daunting. While it is worth noting titles on the history of ecology, there seems little point, for example, in providing an extensive list of works on literary movements such as Romanticism. While there are literally tens of quite adequate biographies of D. H. Lawrence or Mary Shelley and other literary figures, there are only a couple of volumes examining Haeckel (the German scientists who invented the term 'ecology') or seeking to uncover the relationship between the Frankfurt school and contemporary Greens. Library shelves groan with books about Blake but may contain little on the history of ecology. I have also avoided repeating titles that have provided extracts in the main body of the book for obvious reasons of repetition, and I have concentrated on secondary sources, e.g. I have included books on the Frankfurt school rather than by the principal Frankfurt authors such as Marcuse. As well as covering the history of Green ideas, I have included environmental archaeology and history, which, strictly speaking, look at the entire human past in an ecological context rather than simply examining Green movements and themes from history.

Basic definitions

The titles below may help to distinguish ecology as a science from ecology as politics, to define radical environmentalism and to assess whether a distinct Green ideology exists.

O. Owen (1980) *What is Ecology?* Oxford: Oxford University Press. A short and readable introduction to ecology as a science.

J. Porritt (1984) *Seeing Green*, Oxford: Blackwell. A populist and polemical statement of contemporary Green thinking by the former Director of Friends of the Earth UK.

D. Wall and P. Kemp (1990) *A Green Manifesto for the 1990s*, Harmondsworth: Penguin. The authors, who, like Porritt, are members of the UK Green Party, provide a rather more eco-socialist account, emphasizing the utopian aspirations of Green. They include a short history of international Green parties and movements.

M. Bookchin (1989) *Remaking Society*, Montreal: Black Rose. Murray Bookchin, the leading Green anarchist and advocate of social ecology, provides an introduction to his sophisticated reading of Green philosophy.

For an eco-feminist view, amongst many other titles, see M. Mellor (1992) *Breaking the Boundaries: Towards a Feminist Green Socialism*, London: Virago.

For an introduction to deep ecology see A. Naess (1973) 'The shallow and the deep, long-range ecology movement: a summary', *Inquiry* 16(1): 95–100. For a critique see Richard Sylvan's 'A critique of deep ecology', *Radical Philosophy* 40–1 (1984) and, of course. Bookchin's work.

T. O'Riordan (1981) *Environmentalism* London: Pion. An extensive analysis of the environmental movement, distinguishing between radical and reformist strands. O'Riordan touches on the origins of the radical environmental movement and provides an extensive bibliography.

D. Pepper (1990) *The Roots of Modern Environmentalism*, London: Routledge. Pepper contrasts 'ecocentric' and 'technocentric' approaches to nature, examines the intellectual origins of the movement, noting Marxist and anarchist approaches to ecology. He looks at the Romantics, Morris and utopian socialism as sources of modern radical environmentalism.

A. Dobson (1990) *Green Political Thought*, London: Unwin Hyman. An introduction to contemporary Green politics that looks at many important areas of dispute such as the argument around eco-feminism and the deep ecology/environmentalism split.

M. Alihen (1964) *Social Ecology: a Critical Analysis*, New York: Copper Square. First published in 1938, Alihen's essay provides an historical overview of the interplay between ecology and sociology. In particular, it is suggested that ecological social science has often merely borrowed inappropriate metaphors from nature to describe human activity and institutions. Hinting at how sociology might be more effectively 'ecologized', *Social Ecology* helps us to describe the nature of a truly ecological politics.

Broad historical surveys

These works look at attitudes towards the environment over a broad sweep of time.

D. Worster (1991) *Nature's Economy*, Cambridge: Cambridge University Press. Worster's history of the science of ecology traces its origins from before the seventeenth century to the present 'Age of Ecology'. Christian nature philosophers such as John Ray and later Gilbert White, the Romantics, Thoreau and Haeckel are examined. The importance of holism is stressed and a distinction is drawn between 'arcadian' and 'imperial' forms of ecology.

C. Merchant (1980) *The Death of Nature*. San Francisco: Harper & Row. This is an extensive history of changing attitudes towards the natural world, written from an eco-feminist perspective. Merchant illustrates how an ethic of economic progress transformed the Earth from a goddess figure to a resource for exploitation. Like Worster, Merchant places an analysis of ecological ideas in a social and material context.

M. Oelschlager (1991) *The Idea of Wilderness*, New Haven: Yale University Press. Concepts of the wild from 'prehistory to the age of ecology'. A detailed and useful study.

R. Nash (1989) *The Rights of Nature*, Madison: Wisconsin University Press. Nash looks at the history of attitudes to nature, including animal rights.

C. Glacken (1967) *Traces on the Rhodian Shore*, Berkeley: University of California Press. An encyclopaedic survey of perceptions of the environment from classical Greece to the end of the eighteenth century.

M. Berman (1981) *The Reenchantment of the World*, New York: Cornell University Press. Berman examines the evolution of a reductionist world view hostile to the natural environment. Drawing upon the utopian vitalist philosophy of Wilhelm Reich, he suggests an alternative epistemology.

P. Marshall (1992) *Nature's Web*, New York: Simon & Schuster. A broad introductory survey of environmental attitudes, which includes material from outside Europe and North America. Particularly strong on animal rights.

Environmental archaeology

Environmental archaeology, unlike environmental history, let alone Green history, is a well established discipline that places the human past in an ecological context. Environmental archaeology may be distinguished from history by its use of physical and biological evidence rather than the written or spoken word as a source. Increasingly the subjects are likely to overlap as historians use pollen diagrams and excavation data to improve their understanding of the past. The works below provide an introduction to and a taste of this, the original science of the ecological yesterday.

K. Butzer (1982) *Archaeology as Human Ecology*, Cambridge: Cambridge University Press. The classic theoretical account of an ecological human science.

D. Clarke (1976) 'Mesolithic Europe: the economic basis' in G. De G. Sieveking, I. H. Longworth and K. E. Wilson eds, *Problems in Economic and Social Archaeology*. London: Duckworth. An interesting case study which provides some evidence for the existence of an 'original affluent society' in Europe.

J. Evans (1976) *The Environment of early Man in the British Isles*. London: Elek. A basic account of the relations between humanity and nature in Britain from the Palaeolithic onwards.

M. Shackley (1981) *Environmental Archaeology*, London: Allen & Unwin. An introduction to methods.

Specific historical studies

These examine attitudes to the environment and evidence of the existence of early Green or environmental movements during specific and relatively short periods of time. With the exception of our first title, by Debus, these have a strong bias to Britain and North America, so should be supplemented with more general works for a full international perspective.

A. Debus (1978) *Man and Nature in the Renaissance*. Cambridge, Cambridge University Press. A discussion of natural history and the biological sciences.

P. Gould (1988) *Early Green Politics*, Brighton: Harvester. An exciting volume, perhaps the first explicit study of an historical Green movement.

K. Thomas (1984) *Man and the Natural World*, Harmondsworth: Penguin. A detailed investigation of changing attitudes towards nature in Britain from 1500 to 1800. Thomas uses literary and documentary evidence to argue his case that during this period sympathy for nature and other species increased.

L. Marx (1964) *The Machine in the Garden*, London: Oxford University Press. Perhaps a North American equivalent of Keith Thomas's book, Marx's examines the contradiction between the pastoral ideal and technological domination from the perspective of a literary and social historian.

R. Nash (1982) *Wilderness and the American Mind*, New Haven; Yale University Press. An exploration of approaches to 'wilderness' in North America.

S. Hays (1959) *Conservation and the Gospel of Efficiency*, Cambridge, Mass.: Harvard University Press. The principal history of 'progressive conservation' in the United States between 1890 and 1920.

Debates and strands

There are, as we have seen, many different elements in the 'Green Package' and some tensions between apparently contradictory beliefs. The titles below provide an introduction to some areas of controversy.

Anarchism

G. Woodcock (1963) *Anarchism*, Harmondsworth: Penguin.

P. Marshall (1992) *Demanding the Impossible*, London: Harper Collins.

These are the two standard accounts of anarchism, with much relevance to Green history. Accounts of figures including Godwin, Kropotkin, Morris, Goldman, Bookchin and unorthodox political groups from the Anabaptists to the Situationist International.

G. Marcus (1990) *Lipstick Traces*. London: Secker & Warburg. A breathless search for the origins of punk rock in situationism and medieval heresy. Of passing relevance and much entertainment.

N. Cohn (1970) *The Pursuit of the Millenium*, London: Paladin. A broadly unsympathetic portrait of 'revolutionary millenarians and mystical anarchists of the Middle Ages'. Useful background information.

Anthropology

The following shed some light on the discussion around the supposed 'Green' status of so-called 'primitive' peoples. Environmental archaeology (see above) shows the diversity of past societies and provides a new source of evidence on the question.

S. Diamond (1974) *In Search of the Primitive*, New Brunswick, N.J.: Transaction.

J. Donald Hughes (1983) *American Indian Ecology*, El Paso: Texas Western College Press.

J. Lichfield (1992) 'White man speaks with forked tongue', *Independent on Sunday*, 26 April, p. 13.

J. Liedloff (1975) *The Continuum Concept*, London: Duckworth.

Economics

J. Martinez-Alier (1990) *Ecological Economics*, Oxford: Blackwell. An excellent history of ecological energy economics, covering an important but ignored area. This is one of the most important and sophisticated examinations of Green history to be published. It usefully, amongst other questions, distinguishes between Marxian and ecological economics, while suggesting points of common interest between the two types of analysis.

B. Wood (1984) *E. F. Schumacher: his Life and Thought*, New York: Harper & Row. The standard biography of the Green economist by his daughter, it shows the origins (mainly deriving from his time in Burma and resultant encounter with Buddhism) of his ecological awareness.

Far right environmentalism

See works on Malthus and Social Darwinism.

A. Bramwell (1989) *Ecology in the Twentieth Century*, New Haven: Yale University Press. A curious and eclectic study of environmentalist movements and individuals, by a former officer of the Conservative Party's Monday Club (Toczek 1991: 5–9, 35–6; Walker 1977: 126). It emphasizes the far right, for example, looking at 'Green' strands in Nazi Germany, conservative (if not fascist) environmentalists in 1930s Britain such as Rolf Gardiner and Jorian Jenks, while ignoring many radical ecologists, including Aldous Huxley and Gandhi. Bramwell does at least provide a useful bibliography and indicate avenues for further research.

J. L. Finlay (1972) *Social Credit*, Montreal: McGill–Queen's University Press. Background on an otherwise obscure movement for financial reform which had strong links with both ecological groups and the far right in the 1930s.

D. Gasman (1971) *The Scientific Origins of National Socialism*, London: Macdonald. Haeckel, the biologist who coined the term 'ecology', is examined as a Social Darwinist and his 'monist' philosophy suggested as a key source of Nazi doctrine.

Feminism

B. Taylor (1983) *Eve and the New Jerusalem*, London: Virago. An examination of socialism and feminism in the nineteenth century.

The Ecologist, January/February 1992, 22: 1. A special issue on feminism and ecology.

See Merchant's *The Death of Nature* and Mellor's *Breaking the Boundaries*, cited above, along with Sjöö and Mor's *The Great Cosmic Mother*, below.

The Frankfurt School

E. Fromm (1979) *To Have or to Be*, London: Abacus. A popular and polemical account, which draws upon Fromm's heritage as a Frankfurt School psychologist, of a radical philosophy integrating Eastern spirituality, Freudian psychology, Marxism and ecological politics. Useful for assessing the significance (or otherwise) of the school to the modern Green movement, it is often cited by activists (Porritt 1984: 242).

D. Kellner (1984) *Herbert Marcuse and the Crisis of Marxism*, London: Macmillan. An excellent analysis of Marcuse's work, including his views on alienation, the environment and technology.

A. Sohn-Rethel (1978) *Intellectual and Manual Labour: a Critique of Epistemology*, London: Macmillan. A Marxist philosopher close to the School, Sohn-Rethel provides an account critical of reductionist and positive science, parallel to a holistic Green analysis.

Suggested further reading

F. Alford (1985) *Science and the Revenge of Nature: Marcuse and Habermas*, Gainesville: University of Florida Press. Alford examines the heritage of Adorno, Horkheimer and Marcuse in attempting to formulate a non-exploitative attitude towards the natural world. In doing so he provides a critique of Jürgen Habermas, heir to the school, who, he argues, has broken from this concern. Alford provides 'one of the most rigorous analyses to date on the place of science and nature in the Frankfurt School corpus'.

Marxism

T. Benton (1989) 'Marxism and natural limits', *New Left Review* 178. An interesting proposal for the construction of an ecological historical materialism. See subsequent articles in 1990 and 1991 in *New Left Review* that continue this debate.

L. Krader (1979) 'The ethnological notebooks of Karl Marx: a commentary', in S. Diamond (ed.) *Toward a Marxist Anthropology*, New York: Mouton.

H. Parsons (1978) *Marx and Engels on Ecology*, Westport, Conn.: Greenwood Press. Parson brings together a useful collection of extracts from Marx and Engels, exploring their attitudes towards nature, ecology, soil erosion, etc. His analysis of their perspective and suggestion that the former Eastern European regimes protected the environment are open to obvious criticism.

A. Schmidt (1971) *The Concept of Nature in Marx*, London: New Left Books. An Account from a student of the Frankfurt school.

The debate between Marx and the Greens is covered by a large number of authors, including the German Bahro, Bookchin, Enzenburger, Gorz, Marcuse and others.

Paganism and spirituality

M. Sjöö and B. Mor (1987) *The Great Cosmic Mother,* San Francisco: Harper & Row. A sophisticated and encyclopedic study of the goddess in all her aspects. A very well referenced guide to further reading. Sjöö and Mor provide the starting point for investigations of the links between the Green movement and paganism. Highly recommended as a work of feminist research.

R. Hutton (1991) *The Pagan Religions of the Ancient British Isles,* Oxford: Blackwell. A sceptical but sympathetic study based upon recent research.

T. Roszak (1972) *Where the Wasteland Ends*, London: Faber. A polemical and poetic call for a new politics which draws upon the old knowledge of the Romantics, in particular of Blake. It should perhaps be read in tandem with Graves's *The White Goddess*, with which it shares many obvious features. Roszak acts as a very practical link between the contemporary Green movement and its earlier forms. Emotion often looks to displace analysis but Roszak usefully analyses the subjective nature of much supposedly objective Western rational thought.

A. Watts (1958) *Man, Nature, and Woman,* New York: Pantheon. A description of Buddhist and Taoist attempts to see nature as a whole containing the human element.

J. Webb (1976) *The Occult Establishment,* La Salle, Ill.: Library Press.

J. Webb (1981) *The Occult Underground,* La Salle, Ill.: Library Press.

Webb uncovers alternative philosophies from magic to the mystical anarchism of the Diggers of Haight-Ashbury in a well referenced overview of occult movements. Chapter two of *The Occult Underground*, entitled 'Eden's folk', examines 'Green movements' in Britain between 1900 and 1945. These volumes are perhaps best seen as interesting and well researched introductions to an important area largely ignored by historians and sociologists. Webb is strongest when uncovering sources for further work but may be accused of covering too broad a territory to sustain a strong analysis of its full significance.

Utopia

M. Buber (1968) *Paths in Utopia*, London: Routledge. A classic account of utopian thought and practice from Fourier to the Jewish kibbutz movement.

D. Hardy (1979) *Alternative Communities in Nineteenth Century England*, London: Longman. An excellent account of 'practical' utopianism in Britain that identifies religious, socialist and anarchist experiments in community living. A useful introduction to environmentally concerned Victorians, including Ruskin, Morris and Carpenter.

Environmental history

These books look principally at the relationship between human culture and the environment rather than seeking to analyse historical Green or environmental movements.

K. Bailes, ed. (1985) *Environmental History*, Lanham, Md: University Press of America. This series of essays covers topics relevant to a history of Green/environmental attitudes and a study of environmental change in a human context. It includes essays from Worster, Merchant, Crosby, Hughes and other important researchers.

A. Crosby (1987) *Ecological Imperialism*, New York: Cambridge University Press. According to Davis, this account of the 'biological expansions of Europe, 900–1900', 'represents a major first step toward the creation of a comprehensive history of the world environment' (1989: 29). Crosby argues that the animals, plants and, perhaps most significantly, the microbes spread from Europe by colonialists caused greater change than colonial use of military or economic power in transforming the globe.

W. Cronon (1983) *Changes in the Land*, New York: Hill & Wang. An account of how 'ecological imperialism' transformed the environment of New England, looking at the destructive implications of the arrival of a market-based economy. A large introductory chapter 'The view from Walden', reminds us of the social–environmental processes that influenced Thoreau's ecological perspective, centuries before his retreat to the woods.

C. Merchant (1989) *Ecological Revolutions*, Chapel Hill, N.C.: University of North Carolina Press. Like Cronon, Merchant looks at her native New England, examining shifting patterns of land use and thinking from an eco-feminist perspective.

D. Worster, ed. (1988) *The Ends of the Earth*, Cambridge: Cambridge University Press. Essays including 'Doing environmental history' from a number of important researchers in this field.

Bibliographies

E. Davis (1989) *Ecophilosophy*, San Pedro: Miles. Subtitled *A Field Guide to the Literature*, *Ecophilosophy* provides an analytical description of over 300 titles dealing with Green thought. It includes details of many works relevant to 'Green history' and outlines the major contemporary debates of ecologists and radical environmentalists. A very useful tool.

F. Egerton (1977) 'A bibliographical guide to the history of general ecology and population ecology', *History of Science* 15: 189–215.

Bibliography

Adorno, T., and Horkheimer, M. (1972) *Dialectic of Enlightenment*, London: Allen Lane.

Agricola, G. (1556) *De Re Metallica*, transl. H. C. and L. H. Hoover, 1912.

Allaby, M. (1975) 'Ecologists as historians', *Ecologist* 5 (6): 224–5.

Arrhenius, S. (1903) *Lehrbuch der kosmischen Physik*, Liepzig: Hirzel.

St Augustine (1913) *The City of God*, Edinburgh: Clark.

Bacon, F. (1900) *The New Atlantis* (1627) Cambridge: Cambridge University Press.

Bahro, R. (1986) *Building the Green Movement*, London: Heretic.

Balfour, E. (1945) *The Living Soil*, London: Faber.

Ballard, J. G. (1969) *The Disaster Area*, London: Panther.

Bark, W. C. (1958) *Origins of the Medieval World*, Stanford, Col.: Stanford University Press.

Baroja, J. C. (1964) *The World of Witches*, London: Weidefneld & Nicolson.

Barr, J. (1970) 'Calamity supermarket', *Ecologist* 1 (6): 42.

Bauthumley, J. (1678) 'The Light and Dark Sides of God' in N. Smith (ed.) *A Collection of Ranters Writings from the Seventeenth Century*, London: Junction Books.

Bellamy, E. (1922) *Looking Backward* (1889), London: Routledge.

Benton, E. (1989) 'Marxism and natural limits: an ecological critique and reconstruction', *New Left Review* 178: 51–86.

Biehl, J. (1988) 'Deep ignorance', *Greenline* 59: 12–14.

Black, J. (1970) *The Dominion of Man*, Edinburgh: Edinburgh University Press.

Blake, W. (1914) *The Poetical Works of William Blake*, Oxford: Oxford University Press.

Blyth, R. H. (1942) *Zen in English Literature and Oriental Classics*, Tokyo: Hokuseido Press.

Bowra, C. M. (1961) *Greek Lyrical Poetry*, Oxford: Clarendon.

Bramwell, A. (1989) *Ecology in the Twentieth Century*, London: Yale University Press.

Buber, M. (1968) *Paths in Utopia*, London: Routledge.

Bunyard, P., and Morgan-Grenville, F. (1987) *The Green Alternative*, London: Methuen.

Burrows, G. C. (1832) *A Word to the Electors on the unrestricted Use of Modern Machinery*, Norwich.

Callanbach, E. (1975) *Ecotopia*, London: Pluto Press.

Canovan, M. (1977) *G. K. Chesterton: Radical Populist*, New York: Harcourt Brace Jovanovich.

Capra, F. (1983) *The Turning Point*, London: Fontana.

Capra, F., and Spretnak, C. (1986) *Green Politics*, London: Paladin.

Carlyle, J. and T. (1970) *The Collected Letters of Thomas and Jane Welsh Carlyle*, Durham, N.C.: Duke University Press.

Carlyle, T. (1899) 'Signs of the times', in *The Works of Thomas Carlyle in Thirty Volumes* II, London: Chapman & Hall.

Bibliography

Carpenter, E. (1916) *My Days and Dreams*, London: Allen Unwin.

Carpenter, N. (1922) *Guild Socialism*, New York: Appleton.

Carson, R. (1963) *Silent Spring*, London: Hamish Hamilton.

Carver, T. (1983) *Marx and Engels*, Brighton: Wheatsheaf.

Celsus (1980–3) *De medicina*, trans. W. G. Spencer, 3 vols, Loeb Classical Library, London, Heinemann.

Clarke, D. (1976) 'Mesolithic Europe: the economic basis', in G. de G. Sieveking, I. H. Longworth and K. E. Wilson, eds, *Problems in Economic and Social Archaeology*, London: Duckworth.

Cobbett, W. (1926) *Cottage Economy* (1821), London: Peter Davies.

Cohen, M. (1984) *The Pathless Way*, Madison: University of Wisconsin Press.

Conway, A. (1982) *The Principles of the Most Ancient and Modern Philosophy*, The Hague: Martinus Nijhoff.

Conze, E. (1960) *Buddhist Sutras*, Harmondsworth: Penguin.

Cooper, J. F. (1963) *Notions of the Americans* (1828), New York: Ungar.

Darwin, C. (1859) *On the Origin of Species*, London: Murray.

Davis, J. C. (1986) *Fear, Myth and History*, Cambridge: Cambridge University Press.

Day, D. (1989) *The Eco Wars*, London: Harrap.

Debus, A. G. (1978) *Man and Nature in the Renaissance*, Cambridge: Cambridge University Press.

Derrick, C. (1972) *The Delicate Creation*, London: Stacey.

Devall, W., and Sessions, G. (1985) *Deep Ecology*, Salt Lake City: Gibbs Smith.

Diamond, S. (1974) *In Search of the Primitive*, New Brunswick, N.J.: Transaction.

Dobson, A. (1990) *Green Political Thought*, London: Unwin Hyman.

Dumont, R. (1974) *Manifesto of an Alternative Culture*, Paris.

Emerson, R. W. (1883) *Nature: Addresses and Lectures* (1836), Boston: Houghton Mifflin.

Engels, F. (1970) *Socialism: Utopian and Scientific* (1892), Moscow: Progress.

Engels, F. E. (1920) *The Conditions of the Working Class in England* (1844), London: Allen & Unwin.

Evelyn, J. (1930) *Fimifugium* (1661), Oxford: Ashmolean Reprints.

Evelyn, J. (1678) *Sylva*, London.

Finlay, J. L. (1972) *Social Credit*, Montreal: McGill–Queen's University Press.

Forster, E. M. (1951) *Two Cheers for Democracy*, London: Arnold.

Fourier, C. (1972) *Selections from the Work of Fourier* (1822), New York: Gordon.

Fox, G. (1901) *The Journal of George Fox*, London: Friends' Tract Association.

Fox, M. ed. (1987) *Hildegard of Bingen's Book of Divine Works*, Santa Fe: Bear.

Fox, S. (1981) *The American Conservation Movement*, Madison: University of Wisconsin Press.

Frazer, Sir J. (1936) *Aftermath: a Supplement to 'The Golden Bough'*, London: Macmillan.

Freeman, M. (1955) *D. H. Lawrence*, Gainesville, Fla.: University of Florida Press.

Fromm, E. (1961) *Marx's Concept of Man*, New York: Ungar.

Fromm, E. (1976) *To Have or to Be?* London: Abacus.

Gardiner, R. (1949) *Forestry or Famine?* Dublin: Steiner.

Gasman, D. (1971) *The Scientific Origins of National Socialism*, New York: Elsevier.

Gilfillan, S. C. (1965) in *Journal of Occupational Medicine*, 7: 53–60.

Gimbutas, M. (1991) *The Civilisation of the Goddess: the World of old Europe*, San Francisco: Harper & Row.

Glacken, C. (1967) *Traces on the Rhodian Shore*, Berkeley: University of California Press.

Godwin, W. (1793) *Principles of Political Justice*, London.

Goldman, E. (1906–7) *Mother Earth Bulletin* 1: 1.

Goldsmith, E. (1975) 'The fall of the Roman Empire', *Ecologist* 5: 6.

Goldsmith, E. (1988) *The Great U-turn: De-industrializing Society*, Bideford: Green Books.

Goldsmith, E., Allen, R., Allaby, M., Davoll, J., and Lawrence, S. (1972) *Blueprint for Survival*, London: The Ecologist.

Gompertz, L. (1824) *Moral Inquiries on the Situation of Man and Brutes*, London.

Goodison, L. (1992) 'Were the Greeks green?', *Greening the Planet*, 3: 24–5.

Goudie, A. (1981) *The Human Impact*, Oxford: Blackwell.

Gould, P. (1988) *Early Green Politics*, Brighton: Harvester.

Grant, M. (1976) *The Fall of the Roman Empire: a Reappraisal*, New York: Porter.

Bibliography

Graves, R. (1961) *The White Goddess*, London: Faber.

Gray, C. (1974) *Leaving the Twentieth Century: the Incomplete Work of the Situationist International*, London: Free Fall.

Griffin, S. (1978) *Women and Nature*, New York: Harper & Row.

Grundman, R. (1991), 'The ecological challenge to Marxism', *New Left Review* 187: 103–20.

Hardin, G., and Baden, J. (1977) *Managing the Commons*, San Francisco: Freeman.

Hardy, D. (1979) *Alternative Communities in Nineteenth Century England*, London: Longman.

Hardy, T. (1930) 'To the secretary of the Humanitarian League' in F. E. Hardy. *The Later Years of Thomas Hardy*, London: Macmillan.

Harry, M. (1987) 'Attention MOVE! This is America', *Race and Class* 28, 4: 5–28.

Hassall, A. (1861) *Adulterations Detected, or, Plain Instructions for the Discovery of Frauds in Food and Medicine*, London: Longman.

Hawkins, G. (1977) *Beyond Stonehenge*, London: Hutchinson.

Hays, S. P. (1959) *Conversation and the Gospel of Efficiency*, Cambridge, Mass.: Harvard University Press.

Hildegard of Bingen (1990) *Hildegard of Bingen: an Anthology*, London: SPCK.

Horkheimer, M. (1947) *The Eclipse of Reason*, New York: Oxford University Press.

Howard, E. (1898) *Tomorrow*, London: Swan Sonneschein.

Hudson, W. H. (1887) *A Crystal Age*, London: Fisher Unwin.

Hughes, J. Donald (1975a) *Ecology in Ancient Civilizations*, Albuquerque: University of New Mexico Press.

Hughes, J. Donald (1975b) 'Ecology in ancient Greece', *Inquiry* 18 (2): 115–25.

Huxley, A. (1971) *Point Counter Point* (1928), London: Chatto & Windus.

Huxley, A. (1962) *Island*, London: Chatto & Windus.

Huxley, A. (1969) *The Letters of Aldous Huxley*, London: Chatto & Windus.

Illich, I. (1975) *Tools for Conviviality*, London: Fontana.

Jackson, A. (1830) 'Second Annual Message', in J. D. Richardson (ed.) *A Compilation of the Messages and Papers of the Presidents, 1789–1897*, Washington, D.C.

Jones, A. H. M. (1966) *The Decline of the Ancient World*, London: Longman.

Jordan, D. P. (1971) *Gibbon and his Roman Empire*, Urbana: University of Illinois Press.

Jowett, B. ed. (1892) *The Dialogues of Plato* III, London: Oxford University Press.

Kagan, D., ed. (1978) *The End of the Roman Empire*, Lexington, Mass.: Heath.

Kelly, P. (1984) *Fighting for Hope*, London: Chatto & Windus.

Kemp, P. and Wall, D. N. (1990) *A Green Manifesto for the 1990s*, Harmondsworth: Penguin.

Kirkwood, G. M. (1974) *Early Greek Monody*, London: Cornell University Press.

Kobert, R. (1909) in P. Diergart (ed.) *Beiträge aus der Geschichte der Chemie*, Leipzig: Deuticke.

Krader, L. (1979) 'The ethnological notebooks of Karl Marx: a commentary', in S. Diamond (ed.) *Toward a Marxist Anthropology*, New York: Mouton.

Kropotkin, P. (1901) *Fields, Factories and Workshops*, London: Swan Sonnenschein.

Kropotkin, P. (1955) *Mutual Aid* (1914), Boston: Horizon.

Lawrence, D. H. (1936) 'Pan in America', in *Phoenix*, New York: Viking.

Leopold, A. (1949) *A Sand County Almanac*, New York: Oxford University Press.

Lewis, C. S. (1943) *The Abolition of Man*, London: Bles.

Lichfield, J. (1992) 'White man speaks with forked tongue', *Independent on Sunday*, 26 April, p. 13.

Liedloff, J. (1975) *The Continuum Concept*, London: Duckworth.

Linné, C. (1806) *A General System of Nature* I, London: Lackington.

Lot, F. (1966) *The End of the Ancient World and the Beginning of the Middle Ages*, London: Routledge.

Lovelock, J. (1979) *Gaia*, Oxford: Oxford University Press.

Lucretious Carus, T. (1937) *De Rerum Natura*, London: Heinemann.

Lytton, C. (1914) *Prisons and Prisoners*, London: Heinemann.

Macmillen, R. (1967) *Enemies of the Roman Order*, Cambridge, Mass.: Harvard University Press.

Malthus, T. R. (1970) *On the Principles of Population* (1798), Harmondsworth: Penguin.

Marcus, G. (1990) *Lipstick Traces*, London: Secker & Warburg.

Marsh, G. P. (1965) *Man and Nature* (1864), Cambridge, Mass.: Belknap Press.

Martin, P.S. (1973) 'The discovery of America', *Science* 179: 969–74.

Bibliography

Martinez-Alier, J. (1990) *Ecological Economics*, Oxford: Blackwell.

Marx, K. (1977) *Early Writings*, Harmondsworth: Penguin.

Marx, K. (1979) *Capital* I, Harmondsworth: Penguin.

Massingham, H. (1941) *Remembrance*, London: Batsford.

Merchant, C. (1980) *The Death of Nature*, San Francisco: Harper & Row.

Mill, J. S. (1871) *Principles of Political Economy* II (1848) London: Longman.

Morris, B. (1970) 'Ernest Thompson Seton and the origins of the Woodcraft movement, 1925–70', *Journal of Contemporary History* 5 (2): 183–94.

Morris, W. (1907) *News from Nowhere* (1890), London: Longman.

MOVE (1986) 'Letter to friends and supporters from MOVE woman prisoners incarcerated at State Correction Inst.', Muncy, Pa:, February (unpublished MS).

Muir, J. (1901) *Our National Parks*, Boston: Houghton Mifflin.

Mumford, L. (1938) *The Culture of Cities*, London: Secker & Warburg.

Murray, M. (1921) *The Witch Cult in Western Europe*, London: Oxford University Press.

Naess, A. (1973) 'The shallow and the deep, long-range ecology movement: a summary' *Inquiry* 16: 95–100.

Nash, R. (1989) *The Rights of Nature*, Madison: Wisconsin University Press.

Nasr, S. H. (1968) *The Encounter of Man and Nature*, London: Allen & Unwin.

Needham, J. (1956) *Science and Civilisation in China* II, Cambridge: Cambridge University Press.

Nriagu, J. (1983) *Lead and Lead Poisoning in Antiquity*, London: Wiley.

Omvedt, G. (1987) 'India's Green movement', *Race and Class* 28, 4: 29–33.

O'Riordan, T. (1981) *Environmentalism*, Harmondsworth: Penguin.

Orwell, G. (1937) *The Road to Wigan Pier*, London: Gollancz.

Ovid (1921) *Metamorphoses*, Loeb Classical Library, London: Heinemann.

Page, D. (1955) *Sappho and Alcaeus*, Oxford: Clarendon.

Papworth, J. (1971) 'By-election notebook', in *Resurgence* 3 (5): 24–7.

Parsons, H. (1977) *Marx and Engels on Ecology*, Connecticut: Greenwood Press.

Passmore, J. (1974) *Man's Responsibility for Nature*, London: Duckworth.

Paul, L. (1951) *Angry Young Man*, London: Faber.

Pennington, W. (1969) *The History of British Vegetation*, London: English Universities Press.

Penty, A. J. (1937) *Tradition and Modernism in Politics*, London: Sheed & Ward.

Pepper, D. (1990) *The Roots of Modern Environmentalism*, London: Routledge.

Perkins, C. G. (1979) *Herland* (1915), London: Women's Press.

Perowne, S. (1966) *The End of the Roman World*, London: Hodder & Stoughton.

Pinchot, G. (1910) *The Fight for Conservation*, Garden City, N.Y.: Harcourt Brace.

Plato (1974) *Timaeus and Criticus*, Harmondsworth: Penguin.

Pontin, J. (1971) 'A lesson to local councils', *Ecologist* 1(6): 15.

Porritt, J. (1984) *Seeing Green*, Oxford: Blackwell.

Prynn, D. (1972) 'Common Wealth', *Journal of Contemporary History* 7 (1): 169–79.

Prynn, D. (1983) 'The Woodcraft Folk and the Labour movement, 1925–70', *Journal of Contemporary History* 18 (1): 79–95.

Reclus, R. (1873) *The Earth*, London: Chapman & Hall.

Redclift, M. (1986) 'Redefining the environmental "crisis" in the South', in J. Weston (ed.) *Red and Green*, London: Pluto.

Remondon, R. (1984) *La Crise de l'Empire romaine*, Paris: PUF.

Roszak, T. (1972) *Where the Wasteland Ends*, London: Faber.

Roszak, T. (1979) *Person/Planet*, London: Gollancz.

Rumi, Jaluddin (1930) *The Mathnavi of Jaluddin Rumi* IX, Cambridge: Cambridge University Press.

Ruskin, J. (1877) *Unto this Last* (1861), Orpington: Allen.

Russell, D. (1983) *The Religion of the Machine Age*, London: Routledge.

Sahlins, M. (1972) *Stone Age Economics*, Chicago: Aldine Atherton.

Salt, H. (1892) *Animals' Rights connected in Relation to Social Progress*, London: Bell.

Salt, H. (1896) *Percy Bysshe Shelley*, London: Allen & Unwin.

Schimmel, A. (1987) 'Rumi, Jalad al-din', in *The Encyclopedia of Religion*, London: Macmillan.

Schumacher, E. F. (1972) *Small is Beautiful*, London: Blond and Briggs.

Bibliography

Shelley, M. (1969) *Frankenstein* (1831), London: Oxford University Press.

Shelley, P. B. (1965) *The Complete Works of Percy Bysshe Shelley* V, New York: Gordian Press.

Shelley, P. (n.d.) *A Vindication of Natural Diet* (1813), London: Vegetarian Society.

Singer, P. (1976) *Animal Liberation*, New York: New Review.

Sismondi, J. de (1847) *New Principles of Political Economy* (1819), London: Chapman.

Sohn-Rethel, A. (1978) *Intellectual and Manual Labour: a Critique of Epistemology*, London: Macmillan.

Sommervile, M. (1877) *Physical Geography*, London: Murray.

Starhawk (1979) *The Spiral Dance*, New York: Harper & Row.

Stephenson, J. (1989) *Forbidden Land*, Manchester: Manchester University Press.

Stretton, H. (1976) *Capitalism, Socialism and Nature*, Cambridge: Cambridge University Press.

Syer, G. N. (1971) 'Ahead of their time', *Ecologist* 1 (7): 13–14.

Thomas, K. (1984) *Man and the Natural World*, Harmondsworth: Penguin.

Thoreau, H. D. (1964) *Civil Disobedience* (1849), Westwood, N.J.: Revell.

Thoreau, H. D. (1965) *Walden* (1854), London: Dent.

Toczek, N. (1991) *The Bigger Tory Vote*, Stirling: AK Press.

Tolstoy, L. (1901) *The Life and Teachings of Leo Tolstoy*, ed. G. H. Perris, London: Grant Richards.

Trainer, F. E. (1985) *Abandon Affluence!* London: Zed.

Voght, J. (1967) *The Decline of Rome*, London: Weidenfeld & Nicolson.

Walker, A. (1988) *Living by the Word*, London: Women's Press.

Walker, M. (1977) *The National Front*, London: Fontana.

Webb, B. and S. (1941) *Soviet Communism: a New Civilisation* II, London: Longman.

Webb, J. (1976) *The Occult Establishment*, La Salle, Ill.: Library Press.

Webb, J. (1981) *The Occult Underground*, La Salle, Ill.: Library Press.

White, L. (1967) 'The historical roots of our ecological crisis', *Science* 155: 1203–7.

White, G. (1924) *The Natural History of Selborne* (1789), Bristol: Arrowsmith.

Wibaut, Mrs, and Wilkinson, L. G. (1973) *Women in Rebellion* (1900), London: Independent Labour Party.

Wilde, O. (1987) 'The Soul of Man under Socialism', in *De Profundis*, Harmondsworth: Penguin.

Williams, R. (1975) *The Country and the City*, St Albans: Paladin.

Willey, B. (1949) *The Eighteenth Century Background*, London: Chatto & Windus.

Winstanley, G. (1649), 'A declaration from the poor oppressed people of England', in G. Sabine (ed.) *The Works of Gerrard Winstanley*, New York: Russell, 1965.

Wood, B. (1984) *E. F. Schumacher: his Life and Thought*, New York: Harper & Row.

Worster, D. (1991) *Nature's Economy*, Cambridge: Cambridge University Press.

Index

Index

Index

Index

Index